HIGHWAY AND TRANSPORTATION
ENGINEERING AND PLANNING

Frontispiece: A Manchester Metro vehicle at St Peters Square in the city centre

Gavin Macpherson
MSc CEng MICE MIHT MCIT

Highway and transportation engineering and planning

Longman
Scientific &
Technical

Longman Scientific & Technical,
Longman Group UK Limited,
Longman House, Burnt Mill, Harlow,
Essex, CM20 2JE, England
and Associated Companies throughout the world.

Copublished in the United States with
John Wiley & Sons, Inc., 605 Third Avenue, New York
NY 10158

First published 1993

ISBN 0 582 097983

British Library Cataloguing in Publication Data
A CIP record for this book is available from the British Library

Library of Congress Cataloging-in-Publication data
A catalog entry for this title is available from the Library of Congress.

ISBN 0–470–200030 (USA only)

Disc conversion in Times 10/11 point by 8
Printed in Malaysia

Contents

PART I: THE ROLE OF TRANSPORT

Chapter 1: Transport and Society   3

Historical development; transport as a means of extending choice; transport demand and its relationship to land use; traffic assignment and modal choice; congestion and capacity; urban road building and Buchanan; urban traffic control; environmental concern; some transport statistics.

Chapter 2: Transport and Government   19

The constitutional background; local government; Europe; legislation.

PART VI: SCHEME APPRAISAL

PART VII: THE FUTURE

List of Illustrations

Preface

Transport has always been a fundamental part of life in advanced industrial nations, although it has sometimes been taken for granted. In recent years, however, it has come back to the top of the political agenda.

Concern for the environment, whether at the macro level of holes in the ozone layer or global warming, or the micro level of the destruction of habitats, is universal. Transport is a major creator of pollution and one of the major reasons for this concern.

Urban areas are being destroyed by the effects of congestion. But the motor car, the primary cause of the problem, has become an essential part of our way of life. Without it, we could not live in anything like the manner to which we have become accustomed.

Civil engineers have a particularly important role to play in solving transport problems. Some of the biggest construction projects of the 1980s and 90s have been transport projects. Engineers are not only involved in construction but can be expected to take responsibility for planning, management, justification and maintenance as well.

This book aims to set out the procedures and techniques that are required in the planning and implementation of transport systems and to set them in their social, economic and political context.

Gavin Macpherson
Holmfirth, West Yorkshire
September 1992

Acknowledgements

We are grateful to the following for permission to reproduce copyright material:

Bath City Council (Director of Property & Engineering Services) for Figs 22.3 & 22.5; the Controller of Her Majesty's Stationery Office for Figs 6.2(a) & (b) © Crown Copyright, 6.3–6.6, 9.4, 9.5, 11.1, 12.2, 12.20, 12.21, 22.4 & Tables 1.1–1.7, 7.1–7.4, 9.1–9.3, 10.1, 11.1, 25.1 and equation calculations on p. 175; The Institution of Highways & Transportation for Figs 18.1 (Box & Forbes) & 22.6 (Ramsden, Coombe & Bamford); JMP Consultants Ltd for Tables 22.1 & 22.2 (Eastman); Moss Systems Ltd. for Figs 14.1 & 14.2; Thomas Telford Publications and the author, R.D. MacPherson for Figs 22.3 & 22.5 (MacPherson); York College of Further & Higher Education for Figs 22.2(a) & (b).

List of Abbreviations

AADT	Annual Average Daily Traffic
AAHT	Annual Average Hourly Traffic
ACTRA	Advisory Committee on Trunk Road Assessment
COBA	Title of Department of Transport computer program
CAD	Computer Aided Design
CCT	Compulsory Competitive Tendering
DB	Design Bulletin
DRF	Design Reference Flow
DRIVE	Title of European Commission Research Programme
DTp	Department of Transport
EIA	Environmental Impact Assessment
FOSD	Full Overtaking Sight Distance
GIS	Geographical Information System
IHT	Institution of Highways and Transportation
LRT	Light Rapid Transit
LRT	London Regional Transport
LRV	Light Rail Vehicle
MEA	Manual of Environmental Appraisal
MOSS	Title of Highway Design and Ground Modelling computer program
NRTF	National Road Traffic Forecasts

PTA   Passenger Transport Authority

RFC   Reference Flow/Capacity ratio
RHTM   Regional Highway Traffic Model

SACTRA   Standing Committee on Truck Road Assessment
SCOOT   Split Cycle Offset Optimization Technique
SSD   Stopping Sight Distance
SSSI   Sites of Special Scientific Interest

TRANSYT   Traffic Network Study Tool

UDP   Unitary Development Plan
UPM   User Programming Module
UTC   Urban Traffic Control

PART I

THE ROLE OF TRANSPORT

Chapter 1

Transport and Society

Historical development

Transport is the means of moving people and goods between centres of human activity. It is rarely an end in itself, but a means to an end. With very few exceptions, the function of a trip can be defined entirely by the events which preceded it and those which succeed it. Thus, transport is often seen as being subsidiary to other sectors of the economy although it is vital to most of them.

Throughout history, the development of mankind has taken place in parallel with, and been dependent upon, the development of a transport system. Man's dominance is based upon his ability to communicate with his fellow beings. Until the recent invention of telecommunications (themselves the result of a manufacturing process dependent upon transport), he could only do this if he was able to travel.

Since the the earliest times men have wished to trade with their neighbours. This meant the transport of goods. Very complex economic systems, such as the one we enjoy in the United Kingdom, have evolved over thousands of years because of man's need to carry on trade. Even today, without this activity, our society could not exist.[1]

The distribution of the population in the United Kingdom has come about largely as a result of the ease or otherwise of transporting goods. A study of the history of almost any town will show that its relationship to the transport system has been crucial to its development.

London is sited where it is because it is the point at which the road from Rome could cross the estuary of the Thames. Bristol, Hull and Liverpool are all examples of cities which grew at points where land routes reached the coast and safe anchorages were available. Bridges across rivers, or earlier fords, are the basis of a very large number of towns, a fact betrayed by names like Oxford and Cambridge. This is where various roads met, and consequently markets were established. In mediaeval times, and earlier, tradespeople wanted to control the affairs of their markets in the interests of security, cleanliness and, sometimes, preserving monopolies. Most towns will be able to produce early charters granted by the Crown, or a powerful local landowner or the church, granting rights to the leading citizens to control their own affairs. These charters are the origins of local government as we know it today.

The places where people came together for the purpose of carrying on business were convenient locations for local and church administration. These activities, in turn, gave rise to the demand for transport for the people who worked in them needed to be able to get around and needed to buy goods, which had to be brought from the point of manufacture or import or processing to the towns where people lived. Today, it is not unusual for the local authority to be the largest single employer in its area. The journeys to work of its employees can be a major part of the peak-hour traffic flow.

Industrial towns which were established or grew during the industrial revolution may have had their origins far earlier, or may have been dependent on the existence of coal or other mineral deposits. Transport, however, was still important. Newport and Swansea became important because they were situated at the gateway to the mineral-rich interior of South Wales. Blackpool is at the end of the railway line from the industrial north west of England. The railways produced their own towns in which to build and maintain their plant: Swindon, Bletchley and Doncaster were all railway towns. Sometimes the effect on development was negative. In the first half of the nineteenth century, Sheffield can be seen to have been held back by comparison with its northern rivals, Leeds and Manchester. The topography of Sheffield means that the main route out of the city, the Don Valley, points to the north east, whereas the route south was barred by a line of hills until the construction of Bradway Tunnel by the Midland Railway Company. On both the canal and rail systems, Sheffield was at the end of a branch from Rotherham; the trunk routes passed it by. Even the M1 motorway does a great loop to the east of Sheffield, having been originally planned to follow the route of the present M18 to join the Great North Road at Doncaster.

The coming of the mail coaches in the eighteenth century saw a need for staging posts every twenty miles or so along the main trunk routes, thus on the Great North Road, we have Hatfield, Stevenage, Biggleswade, Sandy and so on northwards. Something of a modern equivalent might be said to be found in the way in which airports act as a focus of development, as can be seen at Heathrow, Gatwick, Luton and Birmingham. Something similar will doubtless happen at Stansted.

Modern new towns owe their location to good quality transport links with nearby conurbations. Most were designed with the specific intention of allowing at least some commuting, although they will have some light industry located within their own boundaries. Milton Keynes, which is the newest of all, has good quality rail links with London, which are important to commuters, and it is located adjacent to the M1 motorway, which is important to industry.[2]

Transport as a means of extending choice

The opportunity to exercise choice has come to be seen as a desirable attribute in a developed society. This has been given expression by the Conservative Governments in the UK during the 1980s, the Republican administrations of Reagan and Bush in the United States and the newly emerging democracies of Eastern Europe. The existence of fast, comfortable, safe transport opens up choices to individuals which would not otherwise exist and which we seem to value highly.

The separation of workplace and home – commuting – gives a greater range of job opportunities to individuals living in a given location. Equally, there is a greater choice of home location to individuals working in one place, and this has implications for house prices and therefore the whole economy of areas from which people commute. Given the amount of time that is spent on very long journeys to work, and the real costs involved, it presumably follows that people value this choice quite highly.

But the existence of high commuter flows overloads the transport network during the peak periods when people are travelling to work and travelling home again. Managing traffic flows during the peak period so that delay is minimised without the high economic and environmental cost of urban road building is one of the most complex tasks faced by civil engineers today. If we could solve the problem of congestion at peak times we would have gone a long way to solving not only the transport problems of urban areas, but the social and environmental ones as well.

The cost of transporting goods is eventually reflected in the retail price of those goods. Thus, if costs can be reduced by, for instance, reducing journey times or eliminating double handling, then the price to the customer comes down. That, in turn, increases the range of goods on which the customer can spend his limited resources. The basis of the lower prices for everyday items in large superstores is partly due to the more efficient handling of goods within the store that is available when large quantities are being handled, partly due to the fact that suppliers will give discounts for large orders but also partly due to the more efficient distribution network that is available from centres close to the motorway network. Everyday items are more expensive in Stornoway than they are in Birmingham because, despite the heavily subsidised ferries that serve the Western Isles, there is only a very slow transport system involving high capital costs serving Stornoway.

Transport is inextricably linked with standard of living. If one of our political objectives is to raise living standards – and it is part of the manifesto of every political party with the possible exception of the Greens – then it seems inevitable that demand for transport will rise also.

Transport demand and its relationship to land use

Transport is an essential ingredient in the economic well-being of the country. Its availability gives choice to consumers, but it is also dependent upon the choices that those consumers make, because transport provision cannot be separated from the cost of that provision. The more travellers that there are on a particular link on a network, the less the unit cost of that link is likely to be. Unfortunately, the less the unit cost, the more likely it is that consumers will choose to bear that cost, so we are faced with a perpetual problem of capacity. Providing excess capacity at any part of the network is wasteful of resources. But if demand exceeds capacity then we get congestion and that, in turn, imposes disproportionate costs on travellers and on society as a whole.

An ideal transport network would be one in which supply exactly met demand without there being any spare capacity. Demand, of course, varies with time, so such a network would have to be infinitely flexible, an objective for which we might strive but which we will certainly never reach.

The problem of measuring demand is constantly faced by transport planners. Thirty years ago, the concept of a land use/transportation plan gained credence and several major studies were undertaken. By 1968, planning legislation[3] included requirements on local authorities to produce two levels of development plan: structure plans and local plans. Structure plans were intended to define strategic options over a wide area, and after the reorganisation of local government in 1974, were prepared for each county council.[4] Local plans were prepared in much more detail and might be concerned with areas up to the size of a town. Preparation of these plans involved forecasting future values for a wide range of statistics, including those related to transport.[5]

The potential use for any parcel of land thus came to be defined in a formal document. By the use of its development control powers, a planning authority could ensure, over a period of time, that the plan came to be implemented. It would be possible, by studying existing similar land uses, to calculate the number of trips that would be generated by every area of land, to distribute those trips among all possible destinations, to assign those trips to a network and hence determine the demand for any link on the network (see Figure 1.1). This process, referred to as analytical transport planning, has been fully described by Bruton[6] and others.[7] A large amount of computer power is necessary but it is possible to model the whole of an urban land use/transportation system within a computer.

In practical terms, such a system breaks down at various points. Generation rates cannot be determined simply from land use. A factory may cover a particular area of land but the number, and type, of journeys that will be made by those who work there will depend upon the number of jobs and the amount of production, which may well vary substantially over time according to the prosperity or otherwise of the company. A supermarket may have a particular retail floor area but trips to and from it will vary with the performance of the proprietor in a competitive market.

Box 1.1:

In the past decade Sainsbury's have been very successful, whereas one other national supermarket chain has been dramatically less so. It therefore follows that, for the past decade, Sainsbury's stores have had a far higher trip generation rate than the other chain. Whether this remains true for the next decade is not known, but it is certain that the management of the second chain will be trying to ensure that it doesn't.

So trip generation rates cannot be determined directly from the land use, as they will also depend on the commercial success of the occupant.

At the development control stage, the planning authority will know, in general terms, the future use of the building for which they are being asked to give planning permission. Alternative uses can be categorised by reference to a standard system.[8] Within this system, for instance, A1 refers to retail outlets. This can be further sub-divided into food and non-food retail outlets, but it is not possible to devise a means of discriminating between one occupier and another which would satisfy the absolute need for a planning authority to be seen to be impartial between one commercial organisation and another.

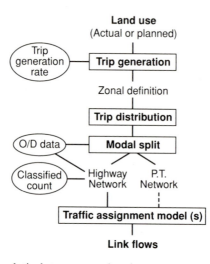

Fig. 1.1   The analytical transport planning process

Traffic assignment and modal choice

A traffic system can best be described as a network, that is, a series of links joining a series of nodes. Trips between pairs of origins and destinations can be assigned to the network, and the number of trips on each link thereby established. The assignment of traffic to a network has become increasingly complicated as individual drivers' perception of congestion has come to influence their decisions on choice of route. As major road building in urban areas has become recognised as being environmentally undesirable and politically unattainable, we are faced with ever increasing demands for the use of scarce road space and consequent congestion. Modern developments in urban transportation planning are aimed at producing a satisfactory model of a congested network.

   To date, it has generally been deemed to be desirable to ensure that travellers have a free choice of mode and route for their journey, limited only by their economic circumstances and some minimal statutory restrictions introduced in the interests of safety. In practice, people have chosen to purchase a car at the first available opportunity. Given a free choice, most people will make a private car 'the second most important purchase of their lives' (after a place to live), a selection seized upon with glee by those charged with marketing motor vehicles.

Having purchased a car, it makes economic sense to use it, as the unit cost per mile will drop the greater the number of miles travelled. Thus, it is very nearly true to say that for most journeys, the modal split between private and public transport is 100%:0% for those journeys where private transport is available. Where private transport is not available, all journeys must be by public transport, but in these circumstances, it is more relevant to consider whether the journey takes place at all.

There are exceptions. Journeys into London by rail or bus do represent a realistic choice which many passengers select. But the same conditions do not occur in other metropolitan areas. One of the major challenges facing transport planners today is to find ways of introducing real choice between public and private transport.

Improvements to urban public transport are presently high on the agenda of transport policy makers and planners. It is, however, too early to say how successful high profile projects such as the Manchester Metro or Sheffield Supertram will be in attracting car-borne commuters, and what level of economic or regulatory inducement will be necessary.

Congestion and capacity

However morally and politically desirable it may be to allow every individual to make their own decision on where to go, when to go and how to go, it is clear that most urban areas in the UK are already suffering from the effects of demand for road space exceeding supply. With predicted increases in car ownership, the situation can only get worse into the future. It is not only an urban problem, as some rural areas, particularly the National Parks such as the Lake District, are increasingly affected by the problems of congestion.

The problem of congestion on a transport network is difficult to predict and to measure, because it depends upon individual traveller's choice of mode, route and time, and this choice will be made in accordance with the traveller's expectation of journey conditions. One person's decision will help to determine the conditions for everyone else. If a train with 500 seats is going to make the journey from Leeds to London, it might just about be possible to squeeze 700 passengers on but no more, although the conditions might be such as to discourage some people from making that journey at that time. The choice of the first 700 has denied the same choice to number 701.

The same is true of every link on every transport network. It has a certain capacity, which exists even though it may not be very easy to define precisely, and the capacity cannot be exceeded, however many individuals want to use their own freedom to take up their own little bit of road space or train seat. With car ownership levels of up to about 10% of the population (the point reached in the UK around 1960), there was room for everyone to make the journeys that they wanted at the time that they wanted. London was a bit different and certain regulations had to be brought in to control traffic flow and parking. The Greater London Development Plan,[9] prepared in 1965 and confirmed in 1976, recognised the need for some traffic restraint. Elsewhere, a major programme of road construction was under way and the perception was that, as people became more prosperous and bought more cars, then the state would build more roads to accommodate them.

Urban road building and Buchanan

The perception of building roads to meet expected traffic flows led to the idea of concentrating traffic flow on major radial routes connected by a ring road or series of ring roads. Coventry, which had suffered massive bomb damage during the Second World War and had to reconstruct most of its central area, provided one of the first complete examples of an Inner Ring Road, with a substantially pedestrianised zone within it. Birmingham and Bristol constructed completed rings whilst Leeds, mindful of the environmental intrusion of a ring road passing close to a major hospital, put some of its (incomplete) inner ring road in cut-and-cover tunnel.

The first suggestion that a policy of urban road building could not be followed to its logical conclusion came in the form of a report to the Minister of Transport by a working party led by Mr (now Professor Sir) Colin Buchanan having the terms of reference 'to study the long term development of roads and traffic in urban areas and their influence on the urban environment'. In his report 'Traffic in Towns',[10] Buchanan pointed out that it simply was not possible to build sufficient roads in urban areas to accommodate forecast traffic levels without entirely reconstructing the town.

In many respects 'Traffic in Towns' was hugely influential, but it is almost as though this central point was missed. For fifteen years after its publication, urban transport planning continued to be based around road construction, albeit with an acceptance that traffic on the new roads would have to be managed and restrained. The introduction of urban traffic control on a wide scale disguised the effects of traffic growth to some extent. So it was not until the late 1980s that the effects predicted by Buchanan really began to be felt in towns in the UK and the need for radical solutions to the problems of congestion became apparent.

Urban traffic control

By the late 1970s emphasis in urban transport planning had switched from increasing the amount of road space in line with the increase in the amount of traffic to seeking to get more traffic on to the roads that had already been built by managing those roads more efficiently. The technology that made this possible was urban traffic control (UTC).

Most major junctions were controlled by traffic signals, and it was already known that, by linking successive signals to a single controller, conditions could be created whereby traffic on a main road travelling at a predetermined speed could always encounter the next set of signals as they turned green. This system is described in detail in Chapter 13. Introducing such linked signals minimised the delay to individual vehicles and hence made more efficient use of road space by ensuring that it wasn't being taken up by standing traffic for more time than was necessary.

The application of pre-programmed microchips to signal controllers meant that far more complicated sequences could be built into the traffic signal system. Computer technology also made possible the modelling of urban networks in such a way that it was possible to investigate different traffic conditions on the computer and devise a series of fixed plans that would optimise signal settings at different times of the day or in response to instructions from a central control room. The necessary software, TRANSYT,[11] was developed at the Transport and Road Research Laboratory, and was rapidly installed in most major urban centres.

One of the problems with TRANSYT is that it is a fixed-plan system of urban traffic control and is based on large quantities of historic data. The signal settings today might have been determined as being optimum two years ago, and traffic conditions might change significantly during that time. A development, which came on stream in the latter part of the 1980s is SCOOT, which is a semi-responsive system. Signal settings are not fixed by predetermined plans brought in at given times of day or in response to instructions from a controller, but respond to actual traffic conditions as they occur.

The effect of the widespread implementation of TRANSYT and SCOOT has been to substantially increase the efficiency of road use in urban areas. This process can continue as UTC is applied in more and more areas, but there is nothing more to be squeezed out of the system in those areas where SCOOT already applies. Once the system is optimised it cannot, by definition, be improved any further.

So by the beginning of the 1990s it can be seen that roads in urban areas during peak periods already handle as much traffic as it is possible for them handle. Further increases in traffic demand cannot be accommodated by getting more vehicles on to the roads. In this situation, what happens is that the peak period extends, so that people change their travel habits in order to try to avoid congested times. In London, the peak already starts at something around seven in the morning and extends until seven in the evening, with very little fall-off in the middle of the day. In other conurbations, peaks are not as extended but the trend is the same. Figure 1.2 shows the typical distribution of traffic by time of day on radial routes in urban areas of different sizes.

Environmental concern

Public concern for the environment has built up during recent years, and road traffic is clearly identified as being one of the villains.[12] Environmental issues are covered in more detail in Chapters 17–20. They have already had a major impact on government policy towards road building, which is, in part, specifically aimed at improving the environment.[13] Almost all road building in urban areas has now been abandoned, whilst the inter-urban programme is heavily biased towards the provision of by-passes which will generate significant environmental benefits, and widening of existing routes, where impact is likely to be minimised.

As we have already seen, a town is a centre of human activity. The transport system links centres of human activity. Therefore the purpose of the (inter-urban) transport system is to provide transport between towns. Journeys begin and end in towns. The development of a transport system that is road-based for the inter-urban part of the journeys but depends on restricting road traffic for the urban part suggests that there is a need for facilities to transfer between modes on the edge of urban areas, and the introduction of park-and-ride schemes indicates that this is beginning to happen. Whether this will lead, in the medium to long term, to the introduction of break-bulk depots for goods traffic, and consequent opportunities for rail carriers to compete for the trunk haul part of a journey, remains to be seen. Alternatively, we may see the development of more out-of-town shopping centres along the lines of Gateshead Metrocentre or Sheffield Meadowhall, where the availability of good motorway access and low-cost, redundant land has determined the location of the facility. Whether this is desirable from the point of view of encouraging dependence on private cars is a matter for future public debate, although both Gateshead and Meadowhall have very good public transport access.

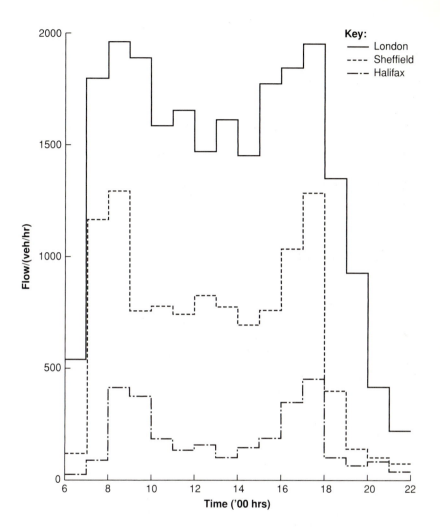

Fig. 1.2 Typical distribution of flow rates by time of day in London, Sheffield and Halifax. **Key:** ———— London, ------ Sheffield, —·—· Halifax

During the period 1990–92 there has been very little increase in car ownership and the general level of traffic, but this has reflected the depth and length of the economic recession through which the country has been passing. Traffic levels are very dependent on economic performance (see Chapter 7), so it is to be expected that they will start to grow again as soon as recession recedes. Current forecasts suggest that total traffic levels will rise by 72% of their 1990 observed values by 2025.[14] How this level of growth is to be reconciled with the

Government's commitment to improve the environment is the major challenge facing engineers and planners today.

Some transport statistics

The significance of transport to the structure of the country is well illustrated by the statistics which are published each year.[15] Apart from the total amount of transport and its cost, the statistics also show a very rapid growth in the transport sector.

In the past ten years total passenger transport has increased by 35.4%, as is shown in Table 1.1.

On average, according to the 1985/86 national travel survey, people travelled an average of 99.5 miles per week, of which 72.2 miles were by car, 7.0 miles by rail and 5.8 miles by bus. Journey purposes are shown in Table 1.2., excluding walking less than 1 mile.

Goods traffic is also on the increase. The total amount carried is shown in Table 1.3.

Table 1.1   Total passenger travel 1980 and 1990 (Source: Reference 15)

Mode	1980	1990	% change
Bus and coach	45	41	−8.8
Car and taxi	395	561	+42.0
Motor cycle	8	7	−12.5
Pedal cycle	5	5	0.0
Rail	35	41	+17.1
Air	3	5	+66.7
Total all modes	491	660	+35.4

Table 1.2   Average weekly travel distance per person by journey purpose (Source: Reference 15)

Journey purpose	Miles
To/from work	22.2
At work	9.5
Education	2.8
Escorting	2.1
Shopping	11.8
Other personal activities	8.6
Social entertainment	28.9
Holidays and day trips	13.7
	99.5

Table 1.3 Freight transport by commodity (Source: Reference 15)

Commodity	tonnes $\times$ 10^9
Agricultural products and livestock	10.9
Foodstuffs	26.4
Solid mineral fuels	12.1
Petroleum products	60.9
Ores and mineral waste	1.8
Metal products	8.9
Minerals and building materials	31.9
Fertilisers	1.9
Chemicals	11.6
Machinery, manufactured goods, miscellaneous	43.2
Total	209.9

Freight transport in 1990 divided by mode was as shown in Table 1.4.

Table 1.4 Freight transport by commodity (Source: Reference 15)

Mode	t.km $\times$ 10^9	%
Road	136.2	63.2
Rail	15.8	7.3
Water	52.5	24.4
Pipeline	11.0	5.1

A fairly cursory glance at these tables allows us to draw some general conclusions regarding transport in the UK:

1. The dominant mode of transport for passenger journeys over a mile is car.
2. The dominance of car travel is becoming greater with the passage of time.
3. There is a slow rise in both the number and proportion of passenger journeys made by rail.
4. The use of buses is declining (despite the deregulation of bus services which was intended to bring service provision more in line with demand).
5. There has been a huge proportionate increase in air travel (for internal transport) but this started from a low base.
6. People do the greatest amount of travelling for reasons associated with leisure.
7. Transport of freight by rail has declined from a once dominant position to the point where it is almost insignificant.

Household expenditure on transport as a proportion of total

expenditure has risen in the ten years between 1979 and 1989, almost entirely due to increased expenditure on car travel. This is shown in Table 1.5.

Public investment in infrastructure is also dominated by road construction as shown in Table 1.6.

Figures shown in Table 1.6 include public investment in rail rolling stock, but do not include any private investment. Private investment in road infrastructure became significant from 1988/89 onwards due to construction of the Queen Elizabeth II Bridge across the Thames at Dartford. In that year it represented about 2.5% of all road investment.

Private investment in rail infrastructure reached £576 million (34.8% of all investment in rail infrastructure) in 1990–91, primarily due to the construction of the Channel Tunnel.

Figures are not available for the amount of private investment in rail rolling stock but it is now anticipated that approximately one-third of the wagon fleet on British Rail is privately owned.

With the Government's stated policy of reducing the number of road accidents by one-third by 2000, attention is being focused on the relative safety records of various modes. This information is provided in Table 1.7.

These figures are sometimes used to make generalised statements of the sort: 'travel by train is 20 times as safe as travel by car', which is true, up to a point. However, to make fair comparisons, it would be

Table 1.5 Household expenditure on transport as proportion of total household expenditure, 1979 and 1989 (Source: Reference 15)

	1979		1989	
	£	%	£	%
All expenditure of which:	94.12	100.0	224.32	100.0
Motor vehicles	10.48	11.1	30.42	13.6
Railway fares	0.56	0.6	0.92	0.4
Bus etc. fares	0.93	1.0	1.33	0.6
Other transport	1.16	1.2	3.10	1.4

Table 1.6 Public investment in transport, $£ \times 10^6$ (Source: Reference 15)

Year	Road	Rail	Ports	Airports
1985/86	2035	567	112	65
1986/87	2120	586	70	66
1987/88	2330	794	65	75
1988/89	2612	811	84	114
1989/90	3032	1119	114	150
1990/91	3704	1478	n/a	173

Table 1.7 Transport casualty rates by mode (Source: Reference 15)

Travellers killed and seriously injured 1980–89, rate per 10^9 passenger km.

Mode	Casualty Rate
Air	0.5
Rail	3.5
Water	*49
Bus and coach	21
Car	67
Motor cycle	2237
Pedal cycle	980
Pedestrian	745

* Safety figures for travel by water are distorted by the *Herald of Free Enterprise* and *Marchioness* disasters in 1987 and 1989 respectively.

necessary to compare rates for similar types of journey. Inter-city journeys, which clock up the maximum number of rail miles, would generate a different casualty rate for car travel, as they would tend to take place on motorways. Most of the injuries which occur to motorcyclists, pedal cyclists and pedestrians are caused when they are in collision with a car. Thus a decision to make a journey by car has signficant implications for the safety of other road users.

The general trend, however, is quite clear from these figures. If the Government is to succeed in bringing down the number of transport casualties, it must look at the area of road transport for all users. If a significant shift to public transport could be achieved, i.e. a reversal of the trends shown in Table 1.1., this would result in an improvement in the transport casualty rate.

References and further reading – Chapter 1

1. For a thorough study of the interaction of transport and the economy generally, see, for instance, Dyos H J & Aldcroft D H *British Transport: An Economic Survey from the Seventeenth Century to the Twentieth.* Leicester University Press. Leicester. 1969.
2. A full description of the current transport system in the UK can be found in Maltby D & White H P *Transport in the United Kingdom.* Macmillan. London. 1982.

3. Town and Country Planning Act 1968.
4. Town and Country Planning Act 1971.
5. Field B & MacGregor B. *Forecasting Techniques in Urban and Regional Planning.* UCL Press. London 1987.
6. Bruton M J *Introduction to Transportation Planning.* 3rd Edition. Hutchinson. London. 1985.
7. Lane R, Powell T J & Prestwood Smith P. *Analytical Transport Planning.* Duckworth. London. 1971.
8. Town and Country Planning Use Classes Order. SI No 1987 764. HMSO. 1987.
9. Greater London Council. *Greater London Development Plan.* GLC. London. 1965.
10. Buchanan Sir Colin (Chairman) – *Traffic in Towns.* The report of the working group. HMSO. London. 1963.
11. Transport and Road Research Laboratory. *TRANSYT user guide.* TRRL Report LR888. Crowthorne. 1980.
12. This Common Inheritance – White Paper on Environmental Policy. Cmnd 1200. HMSO. London. 1990.
13. Roads for Prosperity – the Government's roads programme. White Paper. Cm 693. HMSO. London. 1989.
14. Department of Transport. *Rebasing National Road Traffic.* Forecasts to observed 1990 traffic levels. Circular reference APM 4/1/02/3 dated 1 October 1991.
15. Transport Statistics Great Britain 1991. HMSO.

Chapter 2

Transport and Government

The constitutional background

Transport is very closely regulated in the United Kingdom by both statutory and case law, and it is essential that engineers working in this field understand what they can and cannot do, and the legal basis of their actions.

Like all aspects of English and Scottish law, it has many of its foundations in tradition. Unlike many countries, we do not have a written constitution laid down at some point in history by the founding fathers of the modern state, or by colonial rulers or by political decree. Many would argue that this is a fundamental weakness in the British way of doing things, but it is becoming less important with the growing influence of the European institutions. It remains a fact of life that it is often adequate in law to establish that the principles of a case have been determined by earlier court hearings rather than by an Act of the legislature.

Notwithstanding the significance of case law, Parliament remains sovereign, and it is open to Parliament to change the law if it finds that the courts have come to a decision that it, Parliament, finds unsatisfactory.

Parliament is defined as the House of Commons, the House of Lords and the Monarch acting in concert. By virtue of various pieces of legislation, the House of Commons is by far the most powerful of these three. Its moral authority springs from the fact that it is an elected body, whereas the Lords is created by a mixture of heredity and patronage, and the Monarch is purely hereditary. Pressure has been building up for some years for the replacement of the House of Lords by an elected second chamber with both Labour and Liberal Democrats being broadly in favour. The whole issue of constitutional reform is notoriously complex and one election is unlikely to resolve the issue. The establishment of separate parliaments for Scotland, Wales and possibly the English regions is higher up the priority list for the reformers. At the 1992 election the victorious Conservatives made very plain their opposition to any form of constitutional reform, and the Scottish National Party fared badly. Any change is unlikely at least before the next election. It may well be that the growing importance of European institutions, and the present British Government's commitment to the post-Maastricht blueprint, will lead to domestic reform being overtaken by events.

In order for new legislation to be implemented in the UK, it must be 'read' three times in each of the Commons and the Lords and approved by a simple majority of those present and voting and it must receive the Royal Assent. Legislation can be introduced to Parliament in the form of a bill by the Government, by individual Members (a Private Member's Bill) or by outside agencies such as local authorities (a Private Bill). Major bills introduced by the Government usually go through the Commons first. Detailed work on the bill is undertaken by Committee who will go through it line by line, and revisions are often introduced at this stage. The political composition of the committee will broadly represent the political composition of the House as a whole, not only as between parties but within them as well. Once a bill has passed through the Commons, it goes to the Lords, who repeat the process. The power of the Lords to amend bills is very severely curtailed: they are not allowed to hold up the progress of a Finance Bill, and can only delay other bills for one year.

Political parties are not recognised within the constitution, and it is solely because of the way in which votes are cast at elections that we end up with the vast majority of Members of Parliament being members of one of about half a dozen political parties. What the Constitution does recognise is the existence of the Government and the Opposition. Once an election has taken place, it is the duty of the monarch to appoint a Prime Minister, who must be able to demonstrate that he or she commands a majority in the House of Commons. If an election has produced a result in which one or other of the two largest parties has an overall majority of all the seats in the Commons, then the Leader of that Party is the only viable candidate for the office of Prime Minister (unless the party was itself split so seriously that it could not unite behind its leader). The same conditions do not apply in a so-called 'hung' Parliament in which no party commands an overall majority, or only commands one that is so small that it can be easily out-voted by some combination of the other parties.

There were two occasions in the 1970s when there was no clear candidate for the post of Prime Minister. In February 1974, Edward Heath failed to secure an overall majority for the Conservatives. He remained as Prime Minister after the election, whilst he conducted negotiations with the leader of the Liberal Party, Jeremy Thorpe, to see whether it was possible to form a Conservative–Liberal Coalition Government that would have commanded the support of a majority of the House of Commons. These negotiations failed and Heath resigned. Harold Wilson formed a minority Labour Government which lasted until October 1974, when he called another election and obtained a (small) working majority.

In 1977, Labour's majority in Parliament had been eroded by by-election defeats. Wilson had been succeeded as Prime Minister by James Callaghan. Rather than face defeat in a vote of confidence, which would have meant an immediate election, Callaghan began negotiations with the Liberal Party, led by David Steel. The result was the Lib–Lab Pact, which sustained Callaghan's government in office for a further two years. This was not a coalition government, but an agreement between the two parties by which Liberal MPs would support the Government in any vote of confidence in return for the inclusion of some Liberal measures in the Government's programme. The pact expired at the end of 1978. Callaghan decided not to hold an election immediately but carried on through the so-called 'winter of discontent' until the Spring of 1979. A vote of no confidence was eventually called which Callaghan lost. He resigned immediately, leading to the election of May 1979 and the appointment of Margaret Thatcher as Prime Minister at the head of a majority Conservative administration. The replacement of Margaret Thatcher by John Major in the autumn of 1990 followed from the fact that she ceased to be able to command adequate support from within her own party. If she did not have a majority amongst Conservatives, she certainly did not have a majority in the House, so she could no longer remain Prime Minister.

There is no fixed term for a Parliament within the UK Constitution, but there is a maximum term of five years, after which the Monarch will dissolve Parliament and call an election. There is, however, a strange convention, whereby the Prime Minister can request an early dissolution. This effectively gives the Prime Minister the power to determine the date of the next general election with the proviso that it can be no more than five years after the last one. Thus, the last year or so of a Parliamentary term tends to be characterised by an enormous amount of media speculation which, it can be argued, is very damaging to the national economy.

The general election held on April 9 1992 was widely forecast as being a very close contest. In the event it did not turn out that way and the Conservatives were returned with an overall majority of 21 seats. John Major, who was already Prime Minister, merely continued in that office.

Once the Monarch has appointed the Prime Minister, it is up to him or her to form a Government. This consists of about 100 individuals, about 20 of whom are members of the Cabinet. Ministers are appointed by the Monarch on the recommendation of the Prime Minister. There is no constitutional necessity for them to be Members of Parliament but it would be almost impossible for them to carry out their duties for any length of time unless they were. Amongst senior members of the Cabinet, only the Lord Chancellor is likely to be a member of the House of Lords, the remainder being members of the Commons.

The affairs of state with which we are primarily concerned when considering highways and transport are handled by the Departments of Transport and the Environment, which are currently housed in a trio of ugly tower blocks in Marsham Street, Westminster, close to the Houses of Parliament. In the early 1970s both Departments were combined within a huge Department of the Environment. They were split by Mrs Thatcher when she became Prime Minister in 1979, but for a while transport was in the charge of a Minister of State, who was not a member of the Cabinet. However, transport has become an issue of such political importance that it is inconceivable that it could not now command a seat in the Cabinet. There are certain functions, notably confirmation of orders following public inquiries, that require the two Secretaries of State to act jointly. At the time of writing, the Secretary of State for the Environment is Rt Hon. Michael Howard MP and for Transport is Rt Hon. John MacGregor MP.

Matters relating to transport in Scotland, Northern Ireland and Wales come under the general remit of the Secretaries of State for Scotland, Northern Ireland and Wales. Significant transport offices in the charge of a junior Minister exist in both Edinburgh and Belfast and, to a lesser extent, in Cardiff. These are primarily concerned with administration, which is undertaken by civil servants based in the countries/province concerned, but there are certain detailed differences in the legislation. In England, regional offices of the Departments of Transport and Environment are located in Newcastle upon Tyne, Leeds, Manchester, Birmingham, Nottingham, Bedford, Bristol, Dorking and Kensington. (Addresses of relevant offices are shown at Appendix A).

Local government

The structure of local government in England and Wales was determined by the Local Government Act 1972 as amended by the Local Government Act 1985. In Scotland, the equivalent legislation was the Local Government (Scotland) Act 1983. In Northern Ireland, most functions relevant to transport are undertaken by the Northern Ireland Office, Department of the Environment.

Primary responsibility for trunk roads, main traffic routes in London, airports, railways outside the metropolitan areas and ports and harbours rests with the appropriate central government department. Other transport issues are matters for local authorities.

Local government is subsidiary to central government in every respect, although members of local authorities are directly elected. The authorities, however, are created by Act of Parliament and they can be disbanded by Act of Parliament. The Local Government Act 1985 simply abolished the Greater London Council and the six Metropolitan County Councils in England, transferring their functions to other bodies. All functions and powers of local authorities are laid down by central government, most of them are paid for by the exchequer in London and all other forms of income are controlled centrally. On a day-to-day basis local government has to get the consent of central government on a wide range of issues.

In England, there are six metropolitan areas: West Midlands, Greater Manchester, Merseyside, West Yorkshire, South Yorkshire and Tyne & Wear. These metropolitan counties are subdivided into Districts, each of which has a single unitary authority, responsible for most local services in their area. The Metropolitan District is the highway authority for all roads other than trunk roads in their area and is also responsible for preparation of Unitary Development Plans, development control, traffic management, parking and other environmental functions. Each of the six metropolitan counties has a Passenger Transport Authority (PTA), which is a joint committee of the Districts in its area. Similar arrangements apply to police, fire and ambulance services.

In London, there are 32 London Boroughs plus the Corporation of the City of London, which are, effectively, unitary authorities very similar to metropolitan districts. There is no PTA in London, equivalent powers being exercised by the Secretary of State or, in the case of the police, the Home Secretary. There is also a newly established Traffic Director for London, responsible to the Secretary of State, with London-wide powers for traffic management.

In England, outside London and the six metropolitan areas, and in the whole of Wales, there is a two-tier system of county and district councils. Most transport functions are the responsibility of the county council, who are the highway authority (other than for trunk roads) and have powers to co-ordinate public transport. Districts may become involved through agency arrangements, whereby district councils undertake functions on behalf of the county, or in the areas of development control or parking provision. The Government has announced its intention to move to a unitary system of local government for the whole of Britain, similar to that now operational in metropolitan areas, and a Boundary Commission has been established to give advice on the most appropriate areas for authorities.

In Scotland, the system of local authorities is similar to that in non-metropolitan areas of England except that the upper tier consists of regions rather than counties. Functions related to the water cycle, which are the province of the newly privatised water companies in England and Wales are still part of local government in Scotland. The regional councils are highway authorities (other than for trunk roads) and have powers to co-ordinate public transport similar to those in English counties. There is no PTA in the one Scottish conurbation – Greater Glasgow – so passenger transport functions are undertaken directly by Strathclyde Regional Council. There are three unitary authorities in Scotland covering the three island groups, Orkney, Shetland and Western Isles.

Europe

European institutions are becoming increasingly important in the regulation of transport. Free movement of people and goods across national frontiers is one of the ideals of the original founders of the European Community (EC) and since efficient transport is fundamental to economic well-being, clearly it is essential for moves towards economic union.

A central feature of the EC's policy for regional development is the provision of transport infrastructure. As one of the countries geographically on the edge of Europe, the United Kingdom has benefitted from the application of Regional Development money from Brussels to road, rail and harbour investment. One priority for the EC has been to link all the member states to a Europe-wide network, and the improvement of some trunk routes in the United Kingdom has been possible because they lead, eventually, to Ireland.

There are three levels of European government: the Commission, the Parliament and the Council of Ministers.

The Commission, currently headed by M. Jacques Delors, who is a Frenchman, is the EC's civil service and is located in Brussels. The Commissioners themselves are nominated by the governments of the twelve nation states which make up the Community. The current nominees of the UK are Sir Leon Brittan, a former Conservative Home Secretary and Bruce Millan, a former Labour Secretary of State for Scotland. The President of the Commission, M. Delors, is appointed by the Council of Ministers.

The Commission's function is to propose legislation within certain defined areas for adoption on a European-wide basis, to administer the budget and to ensure that European legislation is implemented by the twelve national governments, if necessary, by reference to the European Court. It employs permanent staff, mostly in Brussels, recruited from among nationals of the Community countries.

The Council of Ministers is actually several Councils, and is composed of the appropriate departmental ministers from the community's twelve national governments. Thus, there will be periodic meetings regarding transport which will be attended by the Secretary of State for Transport (currently Mr John MacGregor) and his counterparts from elsewhere in the Community. Meetings of transport ministers might also involve junior ministers as well as or instead of the Secretary of State. Approximately every three months, there is a European 'summit' attended by the Heads of State of each of the countries to resolve major issues of European politics. In 1991, an important summit was held at Maastricht, Holland, at which very important issues relating to the future of Europe were discussed, including the adoption of majority voting in the Council of Ministers, the establishment of a Central European Bank and the use of a common currency. These issues were considered to be so important that they would have to be ratified by the Parliaments of each of the twelve states, and would not be implemented until they had been so ratified.

In May 1992, Denmark conducted a referendum on the issue of ratifying the Maastricht Agreement and voted, by a very narrow majority, to reject it. Shortly afterwards, Ireland conducted a similar referendum and voted by a large majority to accept it. The policy of the British Government is that it is not necessary to hold a referendum and will seek the approval of the House of Commons in due course. The pace of European reform is therefore very uncertain.

The Presidency of the Council of Ministers rotates among the twelve states, and the United Kingdom succeeded Portugal on 1 July 1992 for a six-month term. An important consideration is the relationship with the newly emerging democracies of Eastern Europe and the former Soviet Union. Unification of Germany has already enlarged the Community and it seems likely that further expansion will take place. In addition, the former European Free Trade Association countries of Austria, Switzerland and Sweden have indicated their desire to join the Community.

The European Parliament is a directly elected body which currently meets in Strasbourg. It has few powers, although it can accept or reject, in its entirety, the Commission's budget.

Legislation

Most legislation relevant to transport can be grouped under the following headings:

1. Transport Acts
2. Highways Acts
3. Road Traffic Acts
4. Town and Country Planning Acts

Two other pieces of legislation with which we are often concerned and which do not fit easily into any of these headings are the Health and Safety at Work etc. Act 1974 and the New Roads and Streetworks Act 1991. In addition, more general legislation such as Local Government Acts, Public Health Acts and Finance Acts can have clauses very relevant to transport and there is a large amount of legislation, some of it dating back a long way, which is concerned with individual projects.

Primary legislation is frequently phrased so as to allow the Secretary of State to 'make regulations' or 'make an order' relating to transport and it is these regulations with which engineers are often concerned when designing, maintaining or administering the transport system. These are published in a variety of forms including Statutory Instruments (SIs) or circulars. These go through a much simpler procedure than bills. SIs are drafted in the Department concerned and 'laid before Parliament' by the relevant Minister. This means that a copy is placed in the House of Commons Library for a defined period. Members of Parliament then have the opportunity to raise matters included in the document either at question time in the House of Commons or directly with the Minister. White Papers, which are

Box 2.1:

An example of the use of SIs might be the control of traffic on a motorway. The primary legislation controlling the power of the Minister to construct highways is the Highways Act 1980, and powers to make regulations controlling traffic are contained in various Road Traffic Acts. The appropriate documents, depending on where the motorway is, are The Motorway Traffic (England and Wales) Regulations 1982 (SI 1982 No 1163) or The Motorway Traffic (Scotland) Regulations 1964 (SI 1964 No 1002) as amended by The Motorway Traffic Amendment (Scotland) Regulations 1968 (SI 1968 No 960) or Motorway Traffic Regulations (NI) 1984 No 160.

Unlike regulations and orders, circulars cannot be used to change the law, but are used to clarify the law and indicate the government's interpretation of the law. It is government policy to seek to reduce the level of casualties resulting from road accidents by one-third of its 1990 level by 2000. One way of doing this is to introduce 20 mph zones. The powers for local authorities to do this already exist, subject to ministerial approval, but local authorities had no experience and could not know what the minister would approve. The result was Circular Roads 4/90.

formal statements of the Government's intentions, are published as SIs and can therefore be debated in Parliament.

Circulars, as the name implies, are simply sent out to interested parties, although they may also be published for sale through government bookshops. They will normally be over the name of a senior civil servant.

Transport Acts[1]

Transport has always been seen as essential to the welfare of the nation and, as such, liable to be regulated by government action. In the eighteenth and early-nineteenth centuries there were a series of Acts to both protect the interests of the shipping companies from the effects of too much competition and to create certain rudimentary safeguards for the safety and welfare of crews and passengers.

The construction of canals and turnpike roads from about 1750 onwards required parliamentary sanction. With the coming of the railways, state involvement increased still further in order to sort out arguments about land acquisition, to ensure safety and to protect customers from the effects of monopoly. Although the early railways were ostensibly private companies in competition with each other, a fixed track system is, by its nature, monopolistic. Large parts of the country were dominated by single companies. No comprehensive legislation was enacted during the time of the construction of the railways, so each company had to promote its own private Act in Parliament. Engineers such as I. K. Brunel had to spend much of their time arguing their case before Parliamentary Committees; Brunel's colleague from the Great Western, Daniel Gooch, was an MP for many years.

By 1921 it had become apparent that there was very little competition between railway companies, and they were grouped into the big four, the London and North Eastern, the London Midland, the Great Western and the Southern, under the terms of the Railways Act 1921. Nationalisation followed with the Transport Act 1947, which established the British Transport Commission to acquire the four railway companies and other transport interests in accordance with the nationalisation policies of the Labour Government elected in 1945.

Since nationalisation, there have been various attempts to sort out the finances of the railways, mostly in response to deficits which were spiralling out of control. The Transport Act 1962 led to the appointment of Dr Richard (Lord) Beeching and a drastic pruning of the network. The Transport Act 1968 endeavoured to define the financial responsibilities of the British Railways Board and to discriminate between those rail services which could, hopefully, be run at a profit and those which were socially necessary. It also established the first Passenger Transport Executives which have since become increasingly important in metropolitan areas. Further changes followed the reform of local government under the Local Government Act 1972 and the subsequent abolition of Metropolitan County Councils under the Local Government Act 1985, leaving the PTAs to function as joint committees of the district councils.

The Conservative Government first elected in 1979 has had 'drawing back the frontiers of the state' as one of its flagship principles. It has steadily withdrawn from direct control of the railways, dividing the network into the profitable inter-city and freight services, leaving the urban networks to Network South East and the PTAs (Passenger Transport Authorities) and making the remainder into regional railways. Each of these is intended to be a separate business and it has more to do with marketing than operation. Privatisation of the profitable parts of the network, with possible franchising for the rest, is a defined objective of the Conservative government re-elected in April 1992.

Highways Acts

The law relating to Highways has its origins in early Mediaeval times. The Statute of Winchester of 1285 decreed that bushes and undergrowth should be cleared on each side of the highway in the interests of safety. The Highways Act 1555, passed in response to the complaints of travellers about the deplorable state of the roads, placed a duty on parish councils to undertake maintenance. This they did by the appointment of a Surveyor who cajoled able-bodied men into giving up to six days free labour per year. The eighteenth century saw the establishment of turnpike trusts to keep the traffic flowing on the major inter-urban routes, and some of the work of the great engineers of the day remains with us to carry traffic in the late twentieth century.

With the coming of the canals and, more particularly, the railways, attention to inter-urban roads tended to lapse. There were a large number of local Acts passed giving particular corporations powers to construct and maintain highways in the great industrial cities that were developing. Some semblance of co-ordination was brought in by the Highways Act 1835 and this remained the basis of much highway law until the first of the two modern statutes, the Highways Act 1959. The last thirty years, however, have seen massive changes in the role of highways in our daily life, and a further Act has been necessary to provide a base from which the many thousands of people concerned with constructing, maintaining and administering highways can work. This is the Highways Act 1980[2,3] and it is currently the most important piece of legislation relevant to Highway Engineering.

The definition of a highway, founded in common law, is a way over which the public has the right to pass and repass. There are different categories of highway depending on whether the public can exercise its right on foot (a footpath), on horseback (a bridleway), on a bicycle (a cycleway) or in a vehicle (a carriageway). A special road is a highway over which the right of the public to pass and repass is limited by ministerial order. Highway authorities are defined in the Act (as subsequently amended) as being The Secretary of State for Transport (or Scotland, Wales or Northern Ireland) acting on behalf of the Crown, the County or Regional Council in non-metropolitan areas, and the Metropolitan Council in metropolitan areas. Ownership of the land over which the highway passes does not necessarily pass to the highway authority although it will normally do so if the highway becomes 'maintainable at the public expense'.

Highways can be created in a number of ways (see Figure 2.1). Major road schemes are promoted by either the local highway authority or the appropriate government department. Such schemes will normally involve land acquisition, which can be either by negotiation or compulsory purchase, and the land so acquired becomes a highway. Minor projects may not involve a full scale land transfer and it may be possible for the highway authority to secure the area it needs by dedication. A scheme involving the widening of a minor road in a rural area might take up a strip of land two or three metres wide and 100 metres long. The owner of the land might well be prepared to dedicate the strip for inclusion into the highway in return for an undertaking from the highway authority that it will build and maintain the fence. A complex legal process can thereby be avoided to the satisfaction of both parties.

Highways can come into being by means of custom and practice. Since the definition of a highway is an area of land over which the public can pass and repass, it has been established that, if the public has been exercising this right without let or hindrance for a significant period of time, the land is deemed to have become a highway. A

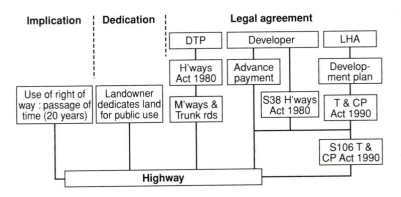

Fig. 2.1 Creation of a public highway

significant period of time is often considered to be 20 years, but it is up to a magistrate's court to determine whether the condition has been met.

A landowner may well wish to allow the public to pass and repass without running the risk of the land becoming a public highway. Developers of modern shopping precincts might need to give the public access to certain parts of their property but would wish to have the freedom to change the route or, indeed, close it altogether. The possibility of creating a highway can be avoided either by closing the route on one day a year (Christmas Day will do) or by erecting a notice to the effect that the public are free to come and go but that the land is NOT designated as a highway, and the right can be withdrawn at any time by the landowner.

In an attempt to remove the potential for argument, the Act places an obligation on the local highway authority to keep and publish a definitive map identifying all public highways in its area.

A common method of creating a new highway is for it to form part of a new development. It will be the subject of an agreement between the developer and the local highway authority under Section 38 of the Highways Act 1980 ('a Section 38 Agreement') by which the developer constructs the highway at his own expense to a specification laid down by the authority. On completion of the development, the authority, on satisfying itself that the highway has been constructed to the required standard, adopts the road which then becomes maintainable at the public expense. There is a different procedure, the Private Streetworks Code, for use in a situation where the road has been constructed at some time in the past without being adopted. The effect is much the same, in that the road becomes a public highway maintainable by the highway authority and the capital cost falls, eventually, on the frontagers.

Under planning law a highway may be created by its inclusion in a Development Plan or by work undertaken as part of an agreement between a developer and the planning authority under section 106 of the Town and Country Planning Act 1990 (A 'Section 106 Agreement') as amended by the Planning and Compensation Act 1991. If the road is a trunk road, the equivalent legislation is Section 278 of the Highways Act 1980.

The New Roads and Streetworks Act 1991 has given the Secretary of State powers to create new highways through private funding on which access will be limited to those who have paid a toll to the company holding the concession. This is granted for a specified period of time. It will be a matter for decision in twenty-five years or so as to whether toll roads created under the 1991 Act revert to being ordinary highways.

Once a public highway has been created or established, the duty of maintainance will normally fall on the landowner. In the case of carriageways this will usually be the highway authority, although there are exceptions. In the case of footways and bridleways it will normally be the landowner, although local authorities have powers to assist under Town and Country Planning and Access to the Countryside legislation.

Road Traffic Acts

Road Traffic Acts are primarily concerned with the management of what happens on the highway. Initially, the main purpose of Parliament was to control the behaviour of drivers of road vehicles in the interests of safety, and measures such as licensing drivers and speed limits were introduced. Control of vehicles on the highway subsequently became important in order to get the most efficient use out a limited amount of roadspace by a rapidly expanding vehicle population, so parking controls, one way streets, traffic signs, bus lanes and restricted turning movements came about. In more recent years, environmental concern has led to regulations allowing compensation of owners of property adversely affected by highway noise.

The arrangement of legislation in this area is still rather complex and there is no single Act which has brought everything under one heading. There is widespread use of regulations and orders. The most important pieces of recent legislation are:

1. The Road Traffic Act 1972, which allowed for licensing of vehicles and drivers, construction and use regulations for vehicles and penalties for driving offences.
2. The Road Traffic Act 1974, which imposes a duty on local authorities to maintain records of road accidents and to promote road safety.
3. The Road Traffic Regulation Act 1984 which includes many of the powers for a local authority to implement traffic management measures relating to speed, movement and parking.
4. Associated regulations such as Local Authorities Traffic Orders (Procedure) (England and Wales) Regulations 1986 and associated documents for Scotland; the Secretary of State's Traffic Orders (Procedures) (England and Wales) Regulations 1986; the London Authorities Traffic Orders (Procedures) Regulations 1986; The Traffic Signs Regulations and General Directions 1981 as amended and the associated Traffic Signs Manual[4] are very important for the implementation of many features of the Road Traffic Acts.
5. The Road Traffic Act 1991 has transferred the management of on-street parking in London from the Metropolitan Police to the London Boroughs, revised definitions of certain driving offences and increased penalties which can be imposed by magistrates.

Recent developments in the application of traffic calming techniques involving the use of road humps[5,6] and 20 mph zones[7] have been dependent upon the issue of regulations by the Department of

Transport giving local highway authorities the necessary powers to carry out the schemes.

Town and Country Planning Acts

There are few aspects of planning that do not have implications for the transport system. The starting point for the planning of transport is frequently statistics related to land use and although the large-scale Land Use Transportation Studies that were undertaken in most urban areas in the late 1960s and early 1970s have not proved as valuable as their proponents had hoped, it is clear that land use planning and transport are closely interwoven.

There are two facets to the planning process that are undertaken by planning officers in local authorities: development plans and development control.

Development plans were originally introduced as a result of the Town and Country Planning Act 1947, which gave local authorities a duty to prepare a plan for their area identifying proposed land uses. This theme was continued with the Town and Country Planning Act 1968 which introduced two levels of plan, the structure plan, identifying proposed land uses over a wide area backed up by a range of economic and social statistics, and the local plan, which was intended to bring together enabling powers for local authorities to undertake work in localised areas. This system was consolidated with the Town and Country Planning Act 1971.

The abolition of the Greater London and Metropolitan County Councils under the Local Government Act 1985 was partly the result of the Government's lack of confidence in the value of conurbation-wide structure plans. The replacement is the Unitary Development Plan (UDP), to be prepared by each London Borough and Metropolitan District under the Town and Country Planning Act 1990. Detailed application of this system to non-metropolitan areas will await the Government's current review of local government boundaries, but it is already clear that a unitary structure will be adopted throughout Britain.

The UDP will include details of road proposals which will be part of the primary route network, together with other transportation proposals that are relevant over the whole of the area covered by the plan or which pass through the area. One of the objectives is to ensure the integration of land use and transportation.

The UDP may also include a broad-brush approach to the management of traffic, including proposals for public transport co-ordination, rail and canal facilities, lorry routes and car parking.

Local plans are prepared by district or unitary councils and must conform to the general policies and proposals of the structure plan/UDP, and may be for quite localised areas. One of their objectives is to bring issues of planning to the notice of the public and a system of public consultation is essential. They provide a detailed basis for development control. With the increased adoption of traffic calming measures as a means of improving the environment of local communities the use of local plans of immediate relevance to the traffic engineer can be expected to grow.

Development control is the system whereby any developer who proposes to carry out work to create or amend a building must, with a few exceptions, submit an application for planning permission to the local planning authority. If there is any possibility that the development will affect a public highway, then the planner will submit the application to the highway authority for comment. A highway engineer will then check to ensure that the proposed development conforms with all the necessary standards for layout, construction, access, parking, and traffic management and will look at the implications for traffic generation and future adoption of highways. In a major development, the highway engineer may well become involved with the developer over possible improvements to the existing network if, without improvement, the network would become overloaded. A wide range of discretion is available to local authorities to reach agreement with developers to allow developments to proceed. A number of major road construction projects have gone ahead through the mechanism of a Section 106 (of the Town and Country Planning Act 1990) or a Section 278 (of the Highways Act 1980, where the works are to be carried out on a trunk road) Agreement whereby the developer, whose site has substantially increased in value with the granting of planning permission, agrees to pay to the local authority a sum of money that can be used for general improvements to the highway network in the area, or for other improvements to services.

If the local planning authority rejects an application, the developer has the right to appeal to the Secretary of State, who may order a public inquiry. The Secretary of State's decision is final unless the applicant can find some way of appealing further to the courts. If a local authority refuses a planning application which is then subsequently granted on appeal, the local authority can find itself ordered to pay the costs of the applicant, and this is a major incentive to local authorities to be wary of turning applications down without very good reason.

New Roads and Street Works Act 1991

This Act had two quite separate purposes: to provide an urgently needed reform of the law covering work in the highway by statutory undertakers, which had previously been the subject of the Public Utilities Street Works Act 1950 (PUSWA) and to provide the necessary legislation for the Government to seek private sector involvement in new road construction schemes. To date, the Birmingham Northern Relief Road and the second Severn crossing are the only private sector schemes to come under the Act, although the Queen Elizabeth II Bridge, Dartford, was constructed by the private sector in return for a toll concession for 25 years.

Parts III and IV of the Act, which replace PUSWA in England and Wales, and Scotland respectively, were originally intended for implementation on 1 July 1992, but this deadline was postponed pending the publication by the Secretary of State of a series of Codes of Practice governing the relationship between local highway authorities and persons/organisations (usually statutory undertakers) carrying out work on the highway and giving guidance on signing and safety measures. The Act finally came into force on 1 January 1993.

The main features of the Act are:

1. Statutory undertakers and other people carrying out work on the highway will be fully responsible for reinstatement of the highway.
2. Local highway authorities will have an overall responsibility for co-ordinating works in the highway so as to minimise disruption. Local highway authorities may require that work is not in progress at certain times.
3. The primary responsibility for ensuring safe and proper practices rests with the organisation promoting the works, who will be specifically required to provide proper signing and guarding.
4. Local highway authorities will, in due course, be required to compile a computerised streetworks register.

Health and Safety at Work, etc. Act 1974

Although not primarily concerned with highways or part of what could be referred to as the law of highways, the Health and Safety at Work Act is so all-embracing that highway engineers should consider it as a part of the legal background to their work.

The Act is concerned with the adoption of safe working practices and sees this as a proper role for managers. In any working situation and, particularly, in a potentially dangerous location such as a highway or a construction site, any person is required to take reasonable action to ensure his own safety and that of anyone for whom he is responsible. The Health and Safety Executive is charged with the responsibility for enforcing the Act and in providing safety guidance.

References and further reading — Chapter 2

1. For more information on Transport Acts, see Hibbs J. *An Introduction to Transport Studies*. 2nd edition. Kogan Page. London. 1988.

2. Highways Open Tech. *Highways Legislation and Administration –
Text 2*. The Highways Act 1980. Bath. 1985.
3. O'Hara R A *Guide to Highway Law for Architects, Engineers,
Surveyors and Contractors*. Spon. London. 1991.
4. Department of Transport. *Traffic Signs Manual*, Chapter 1:
Introduction. HMSO, 1981. Chapter 3: Regulatory Signs. HMSO,
1986. Chapter 4: Warning Signs. HMSO, 1986. Chapter 5: Road
Markings. HMSO, 1985. Chapter 8: Roadworks and Temporary
Situations: HMSO, 1991. Chapter 9: Speed Restrictions. HMSO,
1965.
5. The Highways (Road Humps) Regulations 1990. SI 1990 No 763
and SI 1990 No 1500.
6. The Traffic Signs (Amendment) Regulations 1990 and The Traffic
Signs (Amendment) General Directions 1990. SI 1990 No 704.
7. Department of Transport. *Circular Roads 4/90*. 20 mph Speed Limit
Zones. 1990.

Chapter 3

Transport Expertise

The background of transport professionals

Transport is involved in many spheres of human activity, but it is often at an intermediate stage. As a specialist area of academic study, it tends to be seen as being subordinate to other areas. Civil engineers and planners may well study transportation as a final year option. Business studies students may be offered transport management as a separate stream, alongside retail or public sector management. Separate departments of transport are rare in higher education. The University of Huddersfield has one, and Napier Polytechnic, Edinburgh has a joint department of civil and transportation engineering. Elsewhere, there are substantial units primarily dedicated to research, such as those at Leeds, Newcastle upon Tyne, Oxford, Southampton and Central London, which may be linked to one or more departments.

Transport professionals may thus come from a wide range of academic backgrounds. They may have started off as engineers, economists, geographers, planners, mathematicians or business studies people. They might not have come via a higher education route at all. In further education there are a substantial number of courses, frequently linked to business studies departments, which are aimed at the professional examinations of the Chartered Institute of Transport. There is also provision for gaining the certificates of competence that are required in order to legally run transport services and maintain vehicles. There is thus a particularly good route available through the education system from very lowly levels to professional recognition.

The objectives of transport professionals can be expressed as the provision of the most efficient form of transport available, but this objective is itself subsidiary to other, wider objectives. An understanding of this simple truth is essential if we are to see transport in its proper context and hence to determine its proper role within that context. Seen in isolation, there is a danger that we will regard transport in itself as a good thing, and hence try to increase the size of the transport market when, for social and environmental reasons, we should be trying to do the reverse. Some of the problems of urban congestion could be resolved if the journeys that cause the congestion could be rendered unnecessary by non-transport means. A recent publication by the Association of County Councils[1] has stressed the importance of fully appreciating the transport implications of long term land-use planning.

It is of considerable importance that the right mix of professional expertise is available for any activity. In the case of transport this is perhaps more complex than in other areas because of the mix of skills required. Perhaps the quality of transport provision has suffered in the past, on account of too much fragmentation.

Civil engineers

Traditionally, transport studies were seen as a branch of civil engineering. This was because the major problems associated with the creation of a transport system were those of building the infrastructure. Until the 1840s there was no distinction between the different branches of engineering, other than that between military and civil engineers. The skills that were required to construct turnpike roads, canals, railways and bridges were the skills of masons. These skills had been passed down through the generations from the men who had built the great churches and other public buildings. They were artists who understood the aesthetics and hence, instinctively, the mechanics of shape and form. During the early stages of the industrial revolution they had built the dams to provide power for the mills. The need to pump water out of the mines and speed up the work of the looms prompted them to create engines to harness the power of steam. And it was the ability of the mobile steam engine to replace the horse as the source of motive power that led to a transport revolution.

It was artist/engineers who created the transport systems. Sadly, the breed is almost unknown today, when commercial considerations seem to rule. Perhaps the late Ove Arup was the most recent example and the firm he created comes closest to being the heir to the great Victorian engineers.

In the 1840s there came a split between those who created the track and those who created the engines, the start of the fragmentation which has bedevilled the engineering profession to the present day.

Civil engineers are well qualified to adopt a pragmatic approach to solving problems which can be readily expressed in scientific or mathematical terms. The intensity of modern undergraduate engineering study may have led to shortcomings in the profession's awareness of more subjective issues. Thus, it appears to be difficult for politicians and the public at large to accept that civil engineering expertise is vital to the solution of many of the environmental problems that society faces, including those related to transport.

The strength of mathematics within an engineer's background has meant that he has been able, quite readily, to come to terms with modern technology. Computer aided design is now commonplace in engineer's offices and with the increasing sophistication of both hardware and software making high quality visual images available on screen at the design stage, it may be that future generations will be better able to produce designs of high aesthetic quality.

The qualification system of engineers in the UK is notoriously complicated. It is controlled by the Engineering Council which is itself made up of representatives of more than fifty autonomous institutions. Some reduction in this number is taking place, at least in the civil engineering related area. In 1984 the Institution of Municipal Engineers and Institution of Civil Engineers amalgamated, to become the Institution of Civil Engineers (ICE), and in 1991 the Society of Civil Engineering Technicians joined with ICE. Attempts to arrange a marriage with the Institution of Structural Engineers, however, have so far proved fruitless.

Registration with the Engineering Council can be effected at three levels: Chartered Engineer, Incorporated Engineer and Technician Engineer, dependent upon the level of academic attainment of the individual, the professional training followed and level of responsibility attained. Registration as a Chartered Engineer depends upon having an honours degree approved by the Joint Board of Moderators of the Institutions of Civil and Structural Engineers, and having completed an approved training scheme or gained a large number of years' practical experience. In practice, most people aiming to become Chartered Civil or Structural Engineers will complete an honours degree course in civil engineering at a UK University and then join a major employer on a training scheme that will take between three and five years to complete. They will then have to pass a professional review of their training record, a written test and a formal interview with senior members of the profession.

Incorporated Engineers follow a similar training path but their academic qualification will normally be a BTEC or SCOTVEC Higher

National Diploma or Certificate. Technician Engineers only require a National Diploma or Certificate together with a smaller length of time gaining practical experience. Registration as a Technician Engineer may be a first step towards registration as an Incorporated Engineer.

There is no doubt that the professional bodies carry a lot of weight in determining who comes into the profession and the type of training they receive. This is sometimes said to discriminate against innovation and to isolate many engineers from the public that they serve. The system used for determining qualifications is under continuous review. Concern is frequently expressed about the inappropriate balance between Chartered and Incorporated Engineers and the relative unpopularity of engineering courses among school leavers. The profession is also under considerable pressure following from the gradual erosion of professional values by a commercial ethic. A principal player is the Department of Transport, which has steadily sought to replace contracts in which the Engineer is an arbiter between client and contractor by contracts in which the Engineer is relegated to being simply the designer who seeks to sell his expertise in a commercial market place. The privatisation of the road construction units, the letting of motorway maintenance contracts and the introduction of compulsory competitive tendering to local authority engineering departments are all moves which will fundamentally change the character of engineering.

Managers

Once a transport system has been constructed, it still has to be managed, and the major railway undertakings were among the first huge companies employing more labour than could be controlled by a single entrepreneur. The railways tended to have started life as a consortia of different interests, and an efficient, structured management system was an early requirement. There were particular features necessary to the efficient operation of a transport undertaking, such as time-tabling services and scheduling plant and labour. Co-operation between rival organisations over running powers and pooling ticket receipts developed in the interests of all concerned. Governments took an early interest in regulating services in the interests of safety and fair competition. Thus there grew up a breed of transport managers, initially on the railways and then in bus operations, with somewhat different qualities from those found in manufacturing industry, mineral extraction or public service.

These qualities are today represented by membership of the Chartered Institute of Transport. Membership of the Institute is gained by studying for and sitting the Institute's own exams, although exemptions from some or all of the papers can be obtained by gaining university qualifications.

Architects

Transport buildings are frequently huge in scale: the main hangars at Stansted Airport represent the largest area of roof unsupported by intermediate columns in the UK. They may also be prestigious, such as the Terminal 4 building at Heathrow Airport. So there are outlets for architects in the transport sector, although architecture is not one of the professions usually associated with transport structures. Perhaps this is a pity, for an architect with a professional background involving space, form and aesthetics could perhaps complement the engineer with his pragmatic concern for function. Some would argue (as the Prince of Wales frequently does) that the architectural profession lost its way in the period after the Second World War and became content to produce buildings that were supposed to be functional (although they frequently failed more in this respect than others), depended to a large extent on unitisation, used large quantities of concrete and were very, very boring. Thus, architecture is not strongly represented in the huge transport developments of the period from 1960–90. Engineers held sway, so it is not surprising that we have a motorway system that works well in operational terms but is a gross intrusion on environmental quality.

In recent years, increasing concern for the environment has led to greater attention being given to landscaping transport developments. This can be important in operational terms. The success of a park and ride scheme may well depend upon whether it appears attractive to potential users.

Economists

The social sciences are now coming more to the fore in transport studies. Economic appraisal has become a major factor in transport decision making, frequently based around the principles of cost benefit analysis. This involves allocating values to costs and benefits involved in the transport system, and this is a very difficult process. On the cost side, there is a need to represent the cost to society of using scarce, non-renewable resources, particularly oil. Benefits include the value of time saved by the user and the value of accidents saved, both commodities that can be priced only on the most subjective basis.

By its nature, transport has a major effect on the well-being of the community as a whole, and major parts of it are usually found in the public sector. Thus, some part of the cost does not fall directly on the users but on taxpayers in general. Rigorous economic analysis is necessary to control and justify such public expenditure.

Town planners

As administered in the UK today, planning is essentially a regulatory process carried out by a democratically elected body, the district council. Originally, planning departments were primarily concerned with giving or refusing permission to develop land, and the function was a preliminary stage in the construction process. It was logical that it should fall within the remit of the Borough Engineer who was, until the local government reforms of 1974, a statutory officer, i.e. local authorities were obliged to appoint a properly qualified person to carry out the engineering function.

In fact, practitioners of town planning tended to have backgrounds in architecture and the social sciences rather than in engineering. With the requirement for the drawing up of development plans, which became structure and local plans under the terms of the Town and Country Planning Act 1968 (consolidated in the Town and Country Planning Act 1971) planning became a less reactive activity. With the mushrooming of local authority services after 1974 the activities of the planners became a major influence on the development of transport systems.

It was not long, however, before decline set in. 'The planners' took a lot of the blame for the urban decay that began during the 1960s and has barely been arrested up to the present. This was certainly unfair, for much demolition and reconstruction was promoted by central government and local politicians alike. Buildings of doubtful structural integrity were built because they were cheap and easy to construct. Mrs Thatcher's Government, in office from 1979 until 1990, identified too much planning as being a central factor in our economic ills. In the abolition of the Greater London Council and the six metropolitan county councils in England under the Local Government Act 1985, they destroyed not only a political thorn in their own side but also powerful and influential bodies that were primarily concerned with planning. The *laissez-faire* economic policies with which Thatcher was identified are not compatible with rigid planning of development. But with Thatcher's fall, the pendulum has begun to swing back.

The Town and Country Planning Act 1990 formalises the requirement for unitary development plans in metropolitan areas and the Road Traffic Act 1991 has re-introduced a measure of cross-boundary traffic planning in London. The Government is committed to introducing a unitary system of local government throughout the country, partly with the intention of making responsibility for the planning and the delivery of local services the job of a single body. In transport terms this could lead to the creation of single departments responsible for planning, construction and management of the whole of the transport system. Seen in the light of the Government's commitment to the privatisation of local government services, compulsory competitive tendering and their vision of local authorities as 'enabling bodies' rather than service providers, it is likely that radical change affecting members of all the transport professions is certain in the coming years.

Some sign of the way that things might move are already apparent. For instance, Hertfordshire has removed the traditional title of County Surveyor, replacing it with Director of Transportation. The incumbent, Mr Nigel Knott, is a Chartered Civil Engineer. Next door, in Bedfordshire, a recent reorganisation has brought transportation under the control of the Director of Planning. The adoption of compulsory competitive tendering (CCT) for the work of local authority direct labour organisations has brought the division between client-side and contracting-side functions within the same department. With the extension of CCT to design services, compulsory after 1 April 1994, we can expect the growth of in-house consultancies similar to Sheffield's Department of Design and Building Services. We may also see transportation design departments being hived off to the private sector in the manner already being implemented by Berkshire.

We are going through a period of substantial structural change. Local authorities are looking carefully at their structures, recognising that the professional background of their key officers will have a major influence on the policies that they are able to adopt and implement.

References and further reading — Chapter 3

1. Association of County Councils. *Towards a sustainable transport policy*. ACC Publications. London. 1991.

Chapter 4

Sources of Funding

National and local responsibilities

The operation of transport is, and has always been, split between the public and private sectors. The dividing line varies, and the present policy of the British Government is to increase private involvement, and decrease public. This does not, however, alter the fundamental character of transport. Operations are currently dominated by privately owned vehicles running on publicly owned roads.

Public bodies concerned with transport can normally be defined as being national or local in character, depending on whether they are responsible to central government or to local authorities. Following the transfer of most ports and airports to the private sector, the public sector is primarily concerned with the provision of highways infrastructure, with the financial control of British Rail and with ensuring adequate service levels of public transport .

Highways infrastructure is divided between national and local responsibilities by administrative decree. The construction and maintenance of trunk roads are the responsibility of central government, the Secretary of State being the highway authority. Trunk roads are defined by ministerial order. For all roads that are not trunk roads, the highway authority is a local authority (county, region or metropolitan district according to location). This does not mean, however, that a local authority has much control over the provision of local infrastructure. Local authorities are very closely regulated by central government, from where they obtain the majority of their funds. The UK constitution is such that Parliament is regarded as supreme and a Government which controls Parliament, as the Conservatives have since 1979, is extremely powerful. In the event of conflict developing between national and local authorities, it is certain that Whitehall will win, as the City of Liverpool found out to its cost at the end of the reign of the militant administration of Councillor Derek Hatton.

Local authorities are also empowered to co-ordinate public transport services and to subsidise non-remunerative services. This is a more significant function in the six metropolitan areas of England, where Passenger Transport Executives, originally set up under the Transport Act 1968, still exist, although they are now responsible to a joint committee of the districts in their area. Local authorities are not permitted to run their own public transport services so the former operational companies have been sold off, frequently to management buy-outs.

Ultimate control of all rail services is vested in British Rail, which is responsible to central government. All six PTEs are involved in the operation of rail services, usually through a contractual arrangement with British Rail. Some non-metropolitan authorities, including Nottinghamshire, Leicestershire, Lothian and Strathclyde, are developing structures whereby they can give grant aid to rail services. Initiatives for new and reopened railways comes almost invariably from local authorities. Under the present proposals for privatisation of British Rail,[1] there will continue to be a public sector track authority, but this will be required to operate without subsidy from the exchequer.

Government spending programmes

Transport expenditure is included in the annual review of public expenditure, which is announced by the Government in the autumn of each year. Most of it is included in two bids from Ministers in charge of spending departments: that relating to direct capital expenditure on transport and that relating to local authority expenditure.

The size of the total package of public expenditure is determined by the Chancellor of the Exchequer in consultation with the Prime Minister. It will depend upon the general state of the national economy and the political needs of the government in power.

Once the size of the package has been determined, there follows a series of meetings between the Ministers in charge of spending departments and the Chief Secretary to the Treasury, a member of the Cabinet and the number two finance minister. The existence of these meetings is well publicised but their content remains a closely guarded secret. The judgement that has to be made is an extraordinarily difficult one. Additional expenditure is politically popular and tends to reduce unemployment levels, but it also pushes up the Public Sector Borrowing Requirement (PSBR) which, in the medium to long term, is highly inflationary.

The total amount of public investment in transport between 1985 and 1991 in shown in Table 1.6

When the Secretary of State for Transport knows the amount of resources that he is going to have available for expenditure on trunk roads he will be able to determine which schemes should go ahead. The public expenditure review does not fix expenditure levels precisely, and it is likely to change at short notice in response to changes in world wide economic conditions. It does, however, allow contracts to be signed and schemes to get under way.

Expenditure on most trunk roads comes directly from government funds. This includes maintenance as well as improvements and may be passed to local authorities, acting as agent authority, or to the Department's managing agents for motorways. Although the money comes from the Treasury, there is no relationship between the amount of money received from motor taxation and that paid out on road construction and repair.

Capital expenditure on railways is also subject to central government approval. Despite this, most capital has to be found from within the railways' own resources unless it forms part of the public service obligation grant or comes from PTEs.

Private investment

Until comparatively recently, private investment in transport was limited to ownership of vehicles, rather than track. Some ports, notably Felixstowe, were privately owned and some were owned by British Rail as successors to the private railway companies. Some airports, notably Manchester, Luton, Birmingham and Glasgow (Abbotsinch) were owned by local authorities. Most shipping companies were in private hands, although Scottish ferry services were mostly run by Caledonian MacBrayne, which is a company owned by the government rather than a nationalised industry. The implementation of the government's policies for privatisation have seen British Rail obliged to dispose of its holding in Sealink Ferries (to Stena) and British Airways and the British Airports Authority subject to straightforward flotation on the stock market. Local authorities have been obliged to sell off their airports, as well as their municipal bus undertakings.

Private ownership of freight wagons for running on the railways has always been possible, and, with the rundown of most of the mixed wagon-load traffic, has now become the norm. It is the government's intention to sell off the remainder of rail freight to the private sector. Existing privately owned fleets of wagons may become the model on which privatisation of intercity passenger services is to be based.

The railways have lost much of their parcels traffic to the private sector, and even the high quality Red Star service, carried by express passenger trains, is dependent on private contractors for delivery and collection. There has been enormous growth in the parcels traffic carried by private road vehicles with the expansion of services offered by companies like TNT and Securicor. These companies already offer transfer between road and air on both their international and domestic services, so it will be interesting to see whether, with the privatisation of rail freight services, they also become involved in railways.

The Royal Mail is still a monopoly carrier of the letter post, although the government is believed to be looking at ways of ending the monopoly. They use extensive fleets of lorries together with their own aircraft and, for many of their operations, have to act in direct competition with outside companies.

Private sector investment in rail infrastructure increased very substantially with arrangements for the construction of the Channel Tunnel. Legislation required that a company – Eurotunnel plc – should be established in order to construct and operate the tunnel. This was done, and many members of the public bought shares in the company. In addition, it had to arrange borrowings on the money markets which will be paid back from receipts once the tunnel is open to traffic. Legislation specifically excluded British government investment in the project. With construction nearing completion, attention has turned to the need for a high-speed rail link to London, which is perceived as being unviable without significant financial support from the government.

The New Roads and Street Works Act 1991 opened the way for private capital to be invested in new road building. The Queen Elizabeth II Bridge over the Thames at Dartford was financed by Trafalgar House in return for a concession to charge tolls on both the bridge and the adjacent tunnels, but in most other cases the private sector appears to have been unwilling to take on the full responsibility of expanding Britain's highway network. The exceptions are the second Severn crossing between Bristol and South Wales, and the Birmingham Northern Relief Road.

Substantial private capital has been invested in local highway networks as a result of planning agreements under the terms of Section 106 of the Town and Country Planning Act 1990 (previously Section 52 of the Town and Country Planning Act 1971). Under these agreements, developers undertook to provide improvements to the local highway network in return for being granted planning permission. The justification was that some parts of the profits that accrued from land development could be ploughed back into the community. At the height of the property boom in 1989, as much as 50% of all local authority highway schemes were being paid for in this way, although with the coming of recession, the amount of work in progress has dropped considerably. Section 106 agreements are not limited to provision of highways so, in theory at least, a developer could contribute to the cost of a new bus or rail service.

It is the government's intention to substantially increase the amount of private sector capital invested in railways, although the White Paper 'New opportunities for the railways'[1] suggests that their motives in privatisation are more concerned with the improved management practices which they believe will follow. The main proposals are that:

1. Railfreight and parcels should be transferred in their entirety to the private sector, companies having guaranteed access to the network and appropriate arrangements being made for the transfer of depots and rolling stock.
2. Private sector companies should be given the fullest opportunity to manage and operate existing rail services, grant aided where necessary, through a system of franchising. A new Franchising Authority should be set up to allocate and control franchises.
3. Access to the network for private operators of both freight and passenger services will be overseen by a new Rail Regulatory Body.
4. A new track authority, Railtrack, should be established to have responsibility for all infrastructure and signalling. Railtrack should be funded through levying charges on operators: where subsidy is required on Network South East or Regional Railways services it will be paid to the operators. Railtrack will control train paths subject to overseeing by the Regulator. In the short term, Railtrack will be required to contract out services where this offers value for money and in the long term will itself be privatised.
5. British Rail will continue to operate services in the short term, pending the introduction of a rolling programme of franchises.
6. There will be an opportunity for the private sector to purchase or lease stations.
7. Assistance will be available for existing British Rail staff to purchase legal and financial aid if they are interested in bidding for a franchise.

Local authority funding

Local authorities gain funds for expenditure on transport from four sources:

1. Charges levied on operators for use of facilities or provision of services such as an authority-wide prepaid ticketing scheme.
2. Central government grants for capital expenditure on major projects.
3. Revenue support grants paid by central government as part of the local authority's annual block grant or as part of its Transport Supplementary Grant (TSG).
4. Local taxation, currently the Community Charge or Poll Tax but to be replaced by the Council Tax from April 1993.

Charges levied

It has been the intention of the government since 1979 to reduce the activities of local authorities in providing services very substantially, and to turn them into comparatively small 'enabling bodies'. The work would be undertaken by commercial firms, working under contract.

Many local authorities, however, have sought to frustrate this particular ambition of central government and have established their own organisations, wholly owned by the authority, to compete for contracts and, if successful, to carry out the work. Highway maintenance, waste collection and provision of leisure facilities are all areas where this type of arrangement is common. If the local authority owned contractor is profitable, as it is required to be under the legislation, the profits accrue to the authority.

Capital grants

Local authority expenditure is, by law, divided into capital and revenue items, and different regulations apply in each case. Capital expenditure refers, for the most part, to major schemes and these may attract grant aid from the Exchequer. For instance, improvement schemes to trunk roads carried out as part of a local authority highway improvement will attract grant of 100% of the cost of improving the trunk road. Grant aid to approved capital improvement schemes on principal roads will be 50%. In addition to receiving grant aid, local authorities will have to gain central government's consent to borrow money, if they require to do so, in order to finance their share of the cost.

Revenue grants

The greatest proportion of local authority income comes from central government in the form of revenue grants. These come under two headings: block grant and transport supplementary grant.

Block grant is the money paid by central government to local authorities for running all their services, and they have some discretion over how it is spent. It is based on the Standard Spending Assessment

(SSA), which is a figure computed by the Department of the Environment for each local authority individually. It is the Department's estimate of how much money each local authority should spend in order to provide a standard level of service. It is bitterly disputed by some authorities who maintain that the Government, for political reasons, discriminate against some authorities by allocating a low value of SSA.

Once SSA is determined, all local authorities get a fixed proportion in the form of block grant.

In addition to block grant, local authorities with specific responsibilities for transport also receive finance in the form of Transport Supplementary Grant (TSG). Each year, these authorities submit their Transport Policy and Programme (TPP) document to central government. This sets out their policies and priorities and forms a bid for financial support. Each TPP document is considered individually by central government, and the amount of grant and schemes on which it is to be spent is announced.

Local taxation

The balance of local authority expenditure must be met from local taxation on all people resident and on businesses having premises in the authority's area. Until 1991, this was a property tax called rates. All property was allocated a rateable value, which was based on the rental value, and the local authority fixed a single rate of so much in the pound. The amount payable was then calculated by multiplying the rateable value of an individual property by the rate.

From April 1991, domestic rates were replaced by the community charge, or poll tax, a flat rate tax payable by every individual. Businesses continued to pay a rate fixed by central government and not the local authority. If, in the view of the Government, community charge gets unreasonably high, it can be capped by order of the Secretary of State. A local authority can therefore have a ceiling placed on its turnover and has no alternative but to reduce expenditure. The community charge, however, was enormously unpopular and with effect from April 1993 will be replaced by the Council Tax, which is based on bands of property values.

References and further reading — Chapter 4

1. *New opportunities for the railways: the privatisation of British Rail.* Cmnd 2012. HMSO. London. July 1992.

PART II

TRANSPORT PLANNING

Chapter 5

The Planning Process

Development planning and control

Transport planning is inevitably tied up with land use planning, and it must therefore be related to the statutory framework laid down by the Town and Country Planning Acts.

Legislation relating to most aspects of planning has been overhauled quite recently, and the necessary legislation can be found in the Town and Country Planning Act 1990 as amended by the Planning and Compensation Act 1991. Guidance is issued by the Department of the Environment on the preparation of development plans in the form of Planning Policy Guidance Notes.[1] Counties, which are responsible for preparation of structure plans in non-metropolitan areas, can contribute to regional policy through participation in the Regional Planning Forum.

The major requirements of planning legislation in non-metropolitan areas are as follows:

1. The county council should compile a structure plan for its area. This is a major statement of long term planning policies and proposals for future development and land use in the county. It must indicate how the balance between conservation and development has been struck and how development will be served by transport and other infrastructure. It must be prepared so as to be consistent with national and regional policies laid down by central government, and it forms the framework within which local plans

can be prepared. It is no longer necessary for a Structure Plan to be confirmed by the Secretary of State, although he will monitor progress and may require changes. The life of a Structure Plan is reckoned to be about ten years.

2. The district council should compile a district-wide local plan for its area, consistent with the framework laid down by the county council in the Structure Plan.

3. Mineral local plans and waste local plans should be compiled by the county council to give proposals for mineral extraction and landfill over a 15–20 year programme.

In Metropolitan areas, the structure plan and district-wide local plan are combined into a Unitary Development Plan, but in other respects, the system is similar.

Planning authorities are required to take the following issues into account when preparing development plans:

1. National and regional guidelines.
2. The demand for housing provision.
3. Green belts and conservation of the natural and built environment.
4. The rural economy.
5. The urban economy.
6. Employment generating and wealth creating development.
7. Strategic transport and highway facilities.
8. Mineral workings.
9. Waste treatment and disposal.
10. Land reclamation.
11. Tourism, leisure and recreation.
12. Energy generation

It is intended that major decisions relating to development should be taken at the Development Plan stage, which includes a mandatory programme of public consultation. Applications for planning permission should not normally depart from the provisions of the Development Plans.

Transport Policies and Programmes (TPP)

Every year, each local authority with transport responsibilities (county, metropolitan and London councils in England and Wales, regional and island councils in Scotland) are required to submit a TPP document. This is primarily a bid for funding but it is also required to contain statements of the council's transport policies, details of individual projects and an indication of priority. Major capital schemes will need to be identified in the TPP up to five years in advance of implementation.

The following details, which are included as an example of the procedure, relate to the TPP submission prepared by Kirklees Metropolitan Council, West Yorkshire, for the financial year 1992/93 and submitted to the Department of Transport in July 1991.[2] The document contains the following information:

1. Background information on the local economy, urban renewal and regeneration, industrial, commercial and retail development and tourism.
2. A statement of the Council's policies and objectives.
3. A series of proposals, backed up as necessary by statistics and costings and divided into the following categories:
   - major highways
   - minor highways
   - road safety and accident prevention
   - the environment
   - pedestrians, cycling and the disabled
   - traffic management
   - urban traffic control
   - off street parking
   - street lighting
   - bridges and structures
   - highway maintenance
   - public transport

   The list contains the total of the transport related workload of the authority, not just that expected to attract Transport Supplementary Grant (TSG).
4. Summary of proposals, including priorities for new capital schemes and committed capital and revenue expenditure.
5. Additional resources needed to fund new capital schemes in 1992/3, 1993/4 and 1994/5.
6. Total programme costs amounting to £24.5 million in 1992/3, £29.2 million in 1993/4 and £29.9 million in 1994/5 at out-turn prices, assuming inflation of 8%.
7. The elements of the programme considered eligible for TSG (the bid):

	£'000s	£'000s
MAJOR CAPITAL		
Highways	1682	
Urban traffic control	32	1714
MINOR CAPITAL		
Structural maintenance (bridges)	1650	
Local safety schemes	764	
Structural maintenance (carriageways)	1795	
Traffic management	188	
Street lighting	32	4429

OTHER CAPITAL
Noise insulation works   45
                                                    ―――
                                                    6188

8. TSG Forms:
    1. Summary of expenditure
    2. Highways capital programme: major schemes
    3. Highways capital programme: minor schemes
    4. Structural maintenance programme
    5. Summary of all works

Note the high sums against structural maintenance (bridges) which reflects the urgency of the bridge assessment and strengthening programme made necessary by the heavier lorries which will shortly be permitted throughout the European Community. The local safety schemes figure reflects the policy, over several years, of Kirklees Metropolitan Council to give a high priority to traffic calming and other projects aimed at reducing casualties.

The roads programme

The current programme for improvement and construction of trunk roads in England was defined in a White Paper[3] published in May 1989. This detailed a greatly expanded roads programme for the inter-urban road network, excluding the area within the M25 and other conurbations. Other statements are prepared for roads in Scotland and Wales. The White Paper added some £6 billion to public expenditure.

Prepared in 1988-89, it took account of the very rapid growth in road traffic that had taken place during the 1980s. Traffic overall rose by 35% during the period 1980-89, but it doubled on motorways and increased by 50% on trunk roads. The number of vehicles on British roads increased from 20 million to 23 million.

The White Paper was therefore based on the National Road Traffic Forecasts shown in Table 5.1.

The objectives of the roads programme are stated as follows:

Table 5.1 Forecast % increases in veh.km. 1988–2025 (Source: NRTF reproduced in Reference 3)

	Low	High
Cars	82	134
Light goods vehicles	101	215
Heavy goods vehicles	67	141
Buses and coaches	0	0

1. To assist economic growth by reducing transport costs.
2. To improve the environment by removing through traffic from towns and villages.
3. To enhance road safety.

The White Paper lists a large number of schemes which constitute the roads programme. These include upgrading much of the A1 to motorway standard and widening substantial lengths of the M1, M25 and M6 motorways to dual 4-lane carriageways.

As a White Paper, the roads programme is laid before Parliament and may be the subject of a specific debate. Once Parliament has voted on its acceptance, the need for all the schemes listed is deemed to have been established, although their construction will depend on making the necessary orders to confirm the line of the road and alterations to adjacent side roads, acquiring land and having the necessary funding available.

Since the publication of the White Paper, the economy has been in recession, so traffic levels have not increased as fast as is envisaged in Table 5.1. Many of the schemes listed will certainly be delayed and some might be abandoned altogether before the roads programme is reconstituted by some future White Paper.

Railway planning

Planning of the railways is a matter for the constituent operating units of British Rail, the Passenger Transport Executives and those regional and county authorities who see a role for railways in fulfilling their transport needs or contributing to their economic development.

Major projects, such as the electrification of the East Coast Main Line, were initiated by British Rail and must be paid for out of the railway's resources, i.e. by farepaying passengers. Government approval for expenditure is still required, although this is something that might change if significant private sector involvement occurs in the future. The other BR operating units – Regional Railways, Railfreight and Network SouthEast – are also concerned with long-term planning of their respective sectors.

Smaller electrification projects, such as Leeds–Bradford–Skipton and Ilkley, are planned and promoted by West Yorkshire PTE. Urban rapid transit systems such as Manchester, Sheffield and that planned for West Midlands also originated with the PTEs.

County and regional councils involved in railway projects include Strathclyde, Lothian, Nottinghamshire and Derbyshire.

Private investors, particularly those with substantial wagon fleets of their own, will plan for the future on a strictly commercial basis.

Railway planning, then, is very much an *ad hoc* affair, without any central co-ordination. The national rail network depends very much on how much money can be made available at any given time and the political pressure that can be put in place. There is, for instance, no rolling programme of electrification schemes that would allow contractors to build up teams, assured of work for a reasonable time into the future. Locally, rail services are at the whim of the local authority. While Nottinghamshire, Leicestershire and Derbyshire are pressing ahead with proposals to reopen passenger services on the Robin Hood Line (Nottingham–Mansfield–Retford) and the Ivanhoe Line (Leicester–Ashby–Burton-on-Trent), North Yorkshire are apparently unwilling or unable to prevent a major freight user (British Steel) transferring traffic from rail to road through a National Park. In such a situation, there is little chance of the necessary research being undertaken to determine how best railways could serve the future needs of society and of the environment.

References and further reading – Chapter 5

1. Department of the Environment. Planning policy guidance note no.12. *Development plans and regional planning guidance*. DoE. London. 1992.
2. Kirklees Metropolitan Council. TPP Document for 1992/93. Kirklees MC. Huddersfield. 1991.
3. Roads for prosperity – the Government's roads programme. White Paper. Cm693. HMSO. 1989.

Chapter 6

Data Collection, Storage and Retrieval

Types of transport data

Transport data is required for a wide range of functions concerned with the planning, design and management of the transport system. It can be considered under the following general headings:

1. Journey characteristics: the origin and destination of a trip or potential trip, the time at which it takes place, journey purpose and mode of transport used.
2. Traffic characteristics: the type, speed and turning movements of individual vehicles and traffic streams.
3. Parking studies: the location and duration of parking demand.
4. Accident studies: the location and conditions prevailing at the time of the accident; details of damage and/or injuries received; possible causes.
5. Passenger transport usage, routes, fares, ticketing systems, occupancy and reliability.

These headings are sufficiently general to apply to either passenger or freight transport, and, under most of them, the two would be combined.

Surveys are usually undertaken as an input to the design of facilities, as part of the build-up of a local authority's database used for policy and priority determination or in order to allocate costs and income on a jointly operated venture.

The collection of data is an expensive operation usually involving high manpower requirements. It is therefore imperative that all relevant existing databanks are searched to see whether an individual survey can be avoided. The Department of Transport is concerned to minimise the amount of survey work which involves interviewing members of the public as too much intrusion can be resented and lead to problems of response rate on later surveys. Automatic data collection is possible in many circumstances but the accuracy levels achieved and the range of information gathered are not normally as high as those with manual systems. Greater use is made of automation in the storage and processing of data than its collection. Extensive advice on the practical application of traffic surveys is given in the Traffic Appraisal Manual.[1]

Journey characteristics

It is often necessary to gather extensive data on existing journey patterns in order to forecast future journey patterns and hence to be able to design the transport network that will meet the demand or, if that is not feasible, to determine appropriate measures to manage or restrain the demand.

The first stage in such a study is to define the study area. It is important that this should be sufficiently wide to include all relevant trips, while not introducing lots of extraneous data that will take time to collect and analyse but which will not influence the major conclusions of the study. There is no hard and fast rule about the extent of the study area for a particular scheme. In drawing a cordon and thus defining the study area, the engineer will be concerned to take account of land use patterns, likely journey patterns, the density and capacity of the road and public transport networks and the number of points at which it is possible to cross the cordon. There may be administrative reasons for locating a cordon along a ward, parish or authority boundary. Limiting the number of cordon crossing points will make data collection simpler, so natural or artificial barriers like rivers, railway lines or motorways might make convenient cordons for local studies. If a national zoning system, such as postcodes, is to be used, the cordon must follow zonal boundaries. Figure 6.1 shows an example of a cordon drawn round a small town as part of the preparatory work for constructing a by-pass.

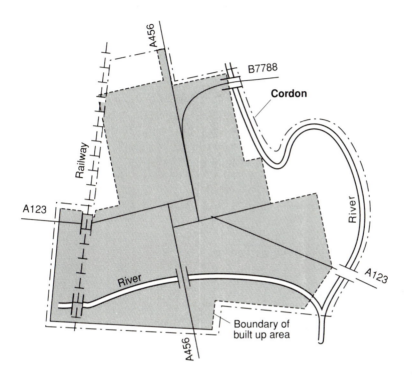

Fig. 6.1 Definition of cordon and study area

In order to analyse origins and destinations it will be necessary to establish a zoning system. The purpose is to aggregate trip ends into geographic areas of origin and destination so that analysis and route assignment can be undertaken by computer.

The Department of Transport maintains a system of zones covering the whole of Great Britain and comprising 78 county or sub-county zones, 447 district zones and 3608 local zones. The system was devised as part of the Regional Highway Traffic Model (RHTM) project in the late 1970s. The RHTM was never completed but the zoning system is available along with other data collected at that time. This system has the advantage of being a national system, thus allowing data to be transferred between one study and another. County and district zonal boundaries were consistent with local government boundaries at the time the system was set up. Sub-county zones are formed because the number of districts within a county exceeded the maximum permitted by the programme. Details, including maps at 1:100 000 scale are available from the Department. The local zones are unlikely to be sufficiently fine to meet the needs of local studies.

An alternative national zoning system is provided by the use of postcodes. There are about 1.5 million postcodes covering the whole of the United Kingdom, which gives a lower size limit of only a few houses. The system is hierarchical in that the first letter or two letters define the postal district and might be shared by several hundred thousand addresses; the postal district is then broken down into a number of areas defined by the number forming the rest of the first half of the code; each area is further divided by the next number, subdivided again by the next letter and finally subdivided by the last letter. Large postal users, which are also likely to be large traffic generators, have a code of their own and can thus be easily identified.

A zoning system can be created using postcodes by having small zones defined by three or four digits from the code close to the centre of the study area and larger zones, further out.

A zoning system defined by postcodes is shown in Figures 6.2(a) and (b). The Ordnance Survey grid reference for each postcode is available from the Office of Population Censuses and Surveys.

For local studies it may not be necessary to have information of trips which do not impinge upon the study area. It may be possible to pick up all relevant trips at the cordon crossing point and this can be taken as the origin. The destination is then either another cordon crossing point or a defined destination within the cordon.

Data suitable for origin and destination analysis can be collected by one of the following techniques:

1. Home interview.
2. Roadside interview.
3. In transit interview.
4. Pedestrian interview.
5. Card distribution and return.
6. Registration number (over limited areas).

Surveys of passenger travel by public transport may be primarily concerned with where the passenger boards and leaves the vehicle being surveyed, in which case the cordon would be drawn tightly round the route itself, cordon crossing points would be stops or stations and all trips would be defined by two cordon crossing points.

Traffic characteristics

Surveyors are likely to be interested in the composition of traffic streams at different times, speeds, journey times, and turning movements in order to carry out analysis of traffic movements and delays for use in programs such as OSCADY (for the design of traffic signal controlled junctions) and for the economic evaluation of

Fig. 6.2(a) National postcode zones. Based on Ordnance Survey map with the consent of the Controller of Her Majesty's Stationery Office. Crown Copyright

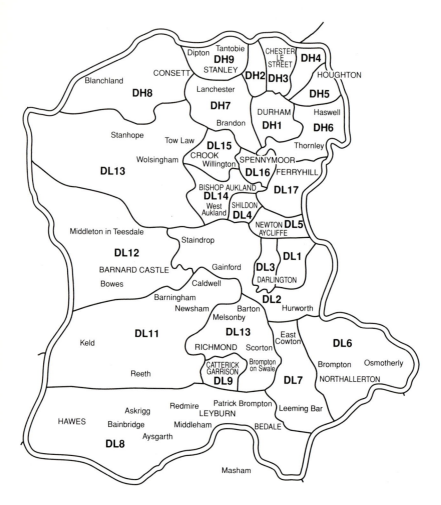

Fig. 6.2(b) Zoning system based on local postcodes for Darlington and Durham. Based on Ordnance Survey map with the consent of the Controller of Her Majesty's Stationery Office. Crown Copyright

schemes. The latter depends heavily on savings in journey time which is computed from the sum of the individual savings. Obviously, a change of speed will have implications for time savings.

Classified counts may be taken at locations within a study area in order to check assignments resulting from origin-destination surveys.

Techniques likely to be used include:

1. Static observer equipped with clipboard or manual tally counter or handheld electronic data capture device and stopwatch.

2. Moving observer similarly equipped.
3. Radar speed meter.
4. Video recorder.
5. Automatic counters.

Parking studies

It is necessary to measure the amount of uncontrolled parking and the effect of parking controls and charges as an input to planning procedures and the management of town centre or rural area parking. Local authorities are obliged to at least break even on their parking operations, and there is therefore a strong incentive to maximise revenue and cater for as much of the demand as is feasible.

Techniques involved are:

1. Desk study using large scale ordnance survey plans (1:1250), structure plans, local plans and unitary development plans.
2. Mobile observer using clipboard or electronic data capture device and stopwatch.
3. Video recording.
4. Card distribution and return.
5. On site interview.

Accident studies

Records of accidents occurring are kept in order to identify locations where remedial treatment might be possible and for the economic evaluation of new road schemes.

Techniques involved are:

1. Transfer of data from police records to build up a continuous and easily accessible database.
2. High speed reaction team on 24-hour standby.

Public transport studies

Surveys of public transport usage are carried out by both local authorities and passenger transport operators as input to the planning and design of bus priority measures, park and ride schemes and interchanges. They are also undertaken to assess the subsidy levels required for operators of unremunerative services and to share out income from common ticketing systems and concessionary fares.

Techniques involved are:

1. Study of ticket issuing records and crew returns.
2. On board interviewer with clipboard or electronic data capture device.
3. Card distribution and return.

Home interviews

Home interview surveys can be used to determine the number of trips of all sorts and by all members of the household. Thus, the children walking to school, the father driving to the office and the mother getting the bus into town will all be covered. Long distance and short distance trips can be included, as can all trips in a given time period, say, 24 hours or 1 week.

The main problem is cost. The survey occupies interviewers and their controllers, often for a large number of evenings. It should, however, attract high quality data covering a wide range of trip patterns, and it is the only way of gathering trip information for unsatisfied demand.

It is first necessary to define the study area in the same way as for any other journey characteristic survey. The attributes which respondents will need in order to take part in the survey must then be defined, e.g. those aged between 18 and 65, those households with access to a car. Finally, it will be necessary to determine the sample size.

Because a 100% sample is likely to be very difficult to achieve (although it is approached with the 10 yearly census home interviews) and would not be significantly more accurate than a smaller sample, it is very rare that home interviews attempt to get a 100% participation rate. The determination of the sample size depends on the accuracy required and a rule of thumb method is shown in the Traffic Appraisal Manual. This is:

$$n = \frac{P(1 - P)N^3}{(E/1.96)^2(N-1) + P(1-P)N^2} \qquad [6.1]$$

where n is required sample size; N is total number of households in area of interest; E is accuracy; P is proportion of households with attributes of interest.

Box 6.1:

Say that we are interested in determining the likelihood of members of a 1-car household living on a particular estate travelling into the town centre by bus. The initial data is as follows:

Number of households on estate	= 1000
Number of households without a car	= 371
Number of household with 1 car	= 603
Number of households with 2 or more cars	= 26

If, say, we require accuracy of +/− 5%, we have to express this as a number of households:

$$E = 5\% \times N \text{ where } N = 1000$$
$$E = 50$$

P is the proportion of households with attribute of interest, i.e. having 1 car

$$P = 603/1000$$
$$= 0.60$$

Substituting in the equation (6.1)

$$n = \frac{P(1 - P)N^3}{(E/1.96)^2(N-1) + P(1-P)N^2}$$

$$= \frac{0.60 \times 0.40 \times 1000^3}{(50/1.96)^2 \times 999 + (0.6 \times 0.4 \times 1000^2)}$$

$$= 41$$

So the surveyor would have to select, at random, 41 out of 1000 households.

We also need to determine the distribution of the sample. Strictly speaking, this should be selected at random. However, if a sample of 200 out of 1000 needs to be selected, a simpler method is to say that 200 out of 1000 is the same as 1 in every 5. A list is drawn up (say, from the electoral register or community charge register) and numbers 1, 6, 11, 16 . . . 996 selected. A slightly more correct approach is to generate a random number between 1 and 5 and use that as the starting point. Most computers will generate a random number.

It will usually be necessary to undertake some preparatory work so as to ensure the co-operation of respondents. This may take the form of publicity in local newspapers or a personalised letter to all potential respondents informing them of the reason for the survey, the method of carrying it out, its timing and the personnel involved.

Selection of interviewers is important, since they will have to be sufficiently knowledgeable about the survey to be able to explain the meaning of the questions. They will have to create confidence and encourage truthfulness. Women are likely to get fuller and more accurate information than men although, in some locations, there may be security concerns. People aged 21–40 will usually be suitable. Care will have to be taken over control of the survey, and adequate checks undertaken to ensure the accuracy of the data collected and recorded. This may take the form of random checks on interviewers' performance, possibly by telephone.

A relatively complicated questionnaire is possible since the interviewer will always be present to explain it to the respondent. Care is still needed to design it carefully, and it should always be piloted before introduction to the full scale survey. Design of questionnaires is a particular skill and the difficulties should not be underestimated. The number of questions and length of time taken to answer them should be chosen so as not run the risk of alienating the respondent by taking up too much of his time. A variation may be to leave a short, simple questionnaire to be filled in by the respondent and then collected by the interviewer who could be available to answer any queries that might have arisen.

Roadside interviews

Data related to origin, destination, vehicle classification, journey purpose and occupancy can be collected by roadside interview. Department of Transport Advice Note TA11/81[2] outlines standard practices and procedures for carrying out roadside interviews.

Locations must be carefully selected bearing in mind the need to be near to the cordon crossing point or screen line which is being surveyed, the safety of both survey staff and motorists and the possibility of delays to traffic. The police should always be consulted and there should always be a police officer present on site directing traffic either into the survey zone or past it.

A standard roadside interview form is shown at Figure 6.3.

The layout and signing of the survey point will vary with the location and the average speed of traffic on the approach. Alternative layouts are shown in section 6 of TA11/81, and an example showing the layout to be used on a single carriageway less than 10m wide with a conveniently located lay-by is at Figure 6.4

At the planning stage it is important to determine the sample size that is going to give a sufficient degree of accuracy when factored up to represent the whole flow. Sample sizes should be as small as possible, consistent with the objectives of the survey in order to minimise the manpower requirement and the delay to traffic.

Interviewer .... 1 [1]	Serial number 2 [ ]	ROADSIDE INTERVIEW Interviewer .... Checked by	Station Number .... 5 [ ]	Date .... d d m m y y 9 [ ]	Time Period Start .... 15 [ ]

VEHICLE TYPE	NO.OF OCCUPANTS		DESTINATION (Full Postal Address)	DESTINATION PURPOSE		ORIGIN (Full Postal Address)	ORIGIN PURPOSE
1 2-wheeled m.v. 2 Car/Taxi 3 Light Goods Veh 4 HGV (2 Axle) 5 HGV (3 Axle) 6 HGV (4 Axle) or more Other .... 18 [1]	1 8 2 9 3 10 4 11 5 12 6 13 7	Address Code P-Postcode G-Grid ref (number) Z-Zone number 23	Firm or House name .... No + street .... Town .... County .... Postcode 24	1 Home 2 Hol Home 3 Work 4 Empl Bus IF VAN [FREIGHT / NON FREIGHT 5 Education 6 Shopping 7 Pers Bus 8 Visit Friends 9 Rec/Leisure 32	Address Code P-Postcode G-Grid ref (number) Z-Zone number 33	Firm or House name .... No + street .... Town .... County .... Postcode 34	1 Home 2 Hol Home 3 Work 4 Empl Bus IF VAN [FREIGHT / NON FREIGHT 5 Education 6 Shopping 7 Pers Bus 8 Visit Friends 9 Rec/Leisure 42
1 2-wheeled m.v. 2 Car/Taxi 3 Light Goods Veh 4 HGV (2 Axle) 5 HGV (3 Axle) 6 HGV (4 Axle) or more Other .... 43 [2]	1 8 2 9 3 10 4 11 5 12 6 13 7	Address Code P-Postcode G-Grid ref (number) Z-Zone number 48	Firm or House name .... No + street .... Town .... County .... Postcode 49	1 Home 2 Hol Home 3 Work 4 Empl Bus IF VAN [FREIGHT / NON FREIGHT 5 Education 6 Shopping 7 Pers Bus 8 Visit Friends 9 Rec/Leisure 57	Address Code P-Postcode G-Grid ref (number) Z-Zone number 57	Firm or House name .... No + street .... Town .... County .... Postcode 59	1 Home 2 Hol Home 3 Work 4 Empl Bus IF VAN [FREIGHT / NON FREIGHT 5 Education 6 Shopping 7 Pers Bus 8 Visit Friends 9 Rec/Leisure 67
1 2-wheeled m.v. 2 Car/Taxi 3 Light Goods Veh 4 HGV (2 Axle) 5 HGV (3 Axle) 6 HGV (4 Axle) or more Other .... 18 [3]	1 8 2 9 3 10 4 11 5 12 6 13 7	Address Code P-Postcode G-Grid ref (number) Z-Zone number 23	Firm or House name .... No + street .... Town .... County .... Postcode 24	1 Home 2 Hol Home 3 Work 4 Empl Bus IF VAN [FREIGHT / NON FREIGHT 5 Education 6 Shopping 7 Pers Bus 8 Visit Friends 9 Rec/Leisure 32	Address Code P-Postcode G-Grid ref (number) Z-Zone number 33	Firm or House name .... No + street .... Town .... County .... Postcode 34	1 Home 2 Hol Home 3 Work 4 Empl Bus IF VAN [FREIGHT / NON FREIGHT 5 Education 6 Shopping 7 Pers Bus 8 Visit Friends 9 Rec/Leisure 42
1 2-wheeled m.v. 2 Car/Taxi 3 Light Goods Veh 4 HGV (2 Axle) 5 HGV (3 Axle) 6 HGV (4 Axle) or more Other .... 43 [4]	1 8 2 9 3 10 4 11 5 12 6 13 7	Address Code P-Postcode G-Grid ref (number) Z-Zone number 48	Firm or House name .... No + street .... Town .... County .... Postcode 49	1 Home 2 Hol Home 3 Work 4 Empl Bus IF VAN [FREIGHT / NON FREIGHT 5 Education 6 Shopping 7 Pers Bus 8 Visit Friends 9 Rec/Leisure 57	Address Code P-Postcode G-Grid ref (number) Z-Zone number 57	Firm or House name .... No + street .... Town .... County .... Postcode 59	1 Home 2 Hol Home 3 Work 4 Empl Bus IF VAN [FREIGHT / NON FREIGHT 5 Education 6 Shopping 7 Pers Bus 8 Visit Friends 9 Rec/Leisure 67

Fig. 6.3 A roadside interview survey form. (Source: Traffic Appraisal Manual)

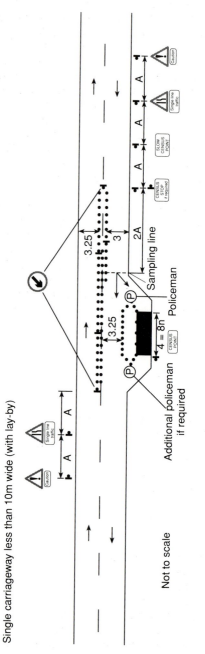

Single carriageway less than 10m wide (with lay-by)

Not to scale

Fig. 6.4   Possible layout for survey site for roadside interview. (Source: TA11/81)

Having determined the required sample size, it is possible to determine the number of interviewers required based on the time taken to complete an interview and the slack time when vehicles are entering or leaving the station. If a total of 1000 interviews are required over 12 hours, i.e. 83 per hour, and each interviewer can complete an interview in 3 minutes, i.e. 20 interviews per hour, we would probably need 4 interviewers, with a fifth on relief duty.

The selection of the sample is a matter for the police officer directing traffic at the survey site. When the previous respondents have moved off, he waits for the stream of vehicles to come along and then waves the second vehicle into the census point. This is to avoid bias due to the front vehicle in a stream being more likely to be a heavy goods vehicle or a bus and therefore unrepresentative of the stream as a whole.

TA11/81 recommends that coding of Origin/Destination (O/D) data should be by means of postcodes but the revised Traffic Appraisal Manual does not agree and suggests instead that six figure ordnance survey grids should be used, making computer analysis simpler.

A classified count should be undertaken at the same time and location as the roadside interview (see Section 6.9) in order to ensure that there is no bias towards particular classifications of vehicles and to determine the precise sample size and total flow.

Manual classified counts

Enumerators can be stationed at the roadside with a clipboard, or tally counter or electronic data capture device and record the number of vehicles in each defined category during a set time period. Turning movements can also be recorded by making one enumerator responsible for traffic entering a junction from each direction and recording each vehicle as turning left, going straight on or turning right.

A standard classification from the Traffic Appraisal Manual is:

1. Pedal cycles.
2. Two wheeled-motor vehicles.
3. Buses and coaches.
4. Light goods vehicles.
5. Heavy goods vehicles with two axles.
6. Heavy vehicles with three axles (rigid).
7. Heavy goods vehicles with four axles (rigid).
8. Heavy goods vevicles with three axles, articulated or draw bar trailer.
9. Heavy good vehicles with five axles or more.

Each of these classifications is illustrated in Figure 6.5. and a suitable data sheet in Figure 6.6.

Increasing use is being made of hand-held data capture devices for recording traffic flows. Several manufacturers produce software for application on hardware such as a Psion Organiser. The Institute for Transport Studies at Leeds University has published a comparison of various devices.[3] Manufacturers are continuously working to introduce greater sophistication and, as with anything dependent on microchip technology, any purchase will rapidly be superseded. Capital constraints, however, are not high, with hardware being available from about £250 (1992 prices). The highest out-going will be the cost of training operators, although this is likely to reduce as these topics find their way into standard educational courses for technicians.

The main advantage in the use of electronic data loggers is in the downloading of raw data for computer analysis and presentation. With straightforward classified counts this is likely to involve the physical removal of a data pack from the logger and its insertion in the office based computer. In more complex arrangements on permanent survey sites, the same function can be undertaken automatically by a fixed link or landline using the Telecom network.

Data loggers are equipped with an inbuilt time clock so the time element of the data can be recorded with greater precision and accuracy than using a stopwatch.

There is no evidence to suggest that use of dataloggers is more or less accurate than pencil and paper methods, although it does avoid the possibility of error at the data transmission stage. The accuracy of classified counting is not all that good; 95% confidence levels of +/− 10% for cars and taxis and +/− 18% for all goods vehicles have been reported.

Registration number surveys

A useful means of recording traffic passing through a relatively small area is a registration number survey. Observers are posted at all the entrances and exits to the area being surveyed. From the time the survey is started to the time it is finished, observers note a part of the registration number of every vehicle that passes, noting whether it is entering or leaving the study area.

It is not necessary to record the whole of the number: the figures and year letter will do. So A123 ABC becomes A123 and ABC 123Y becomes 123Y. There are 21 possible year letters in mainland British registrations (I, O, Q, U and Z are not used) so the system will identify a unique vehicle in about 21,000. Overseas registrations and those from Northern Ireland can be recorded in full but there are

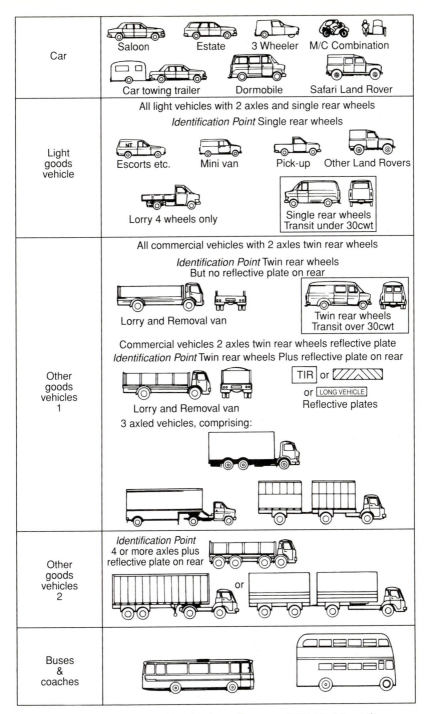

Fig. 6.5 Standard vehicle classifications (Source: COBA9 Manual)

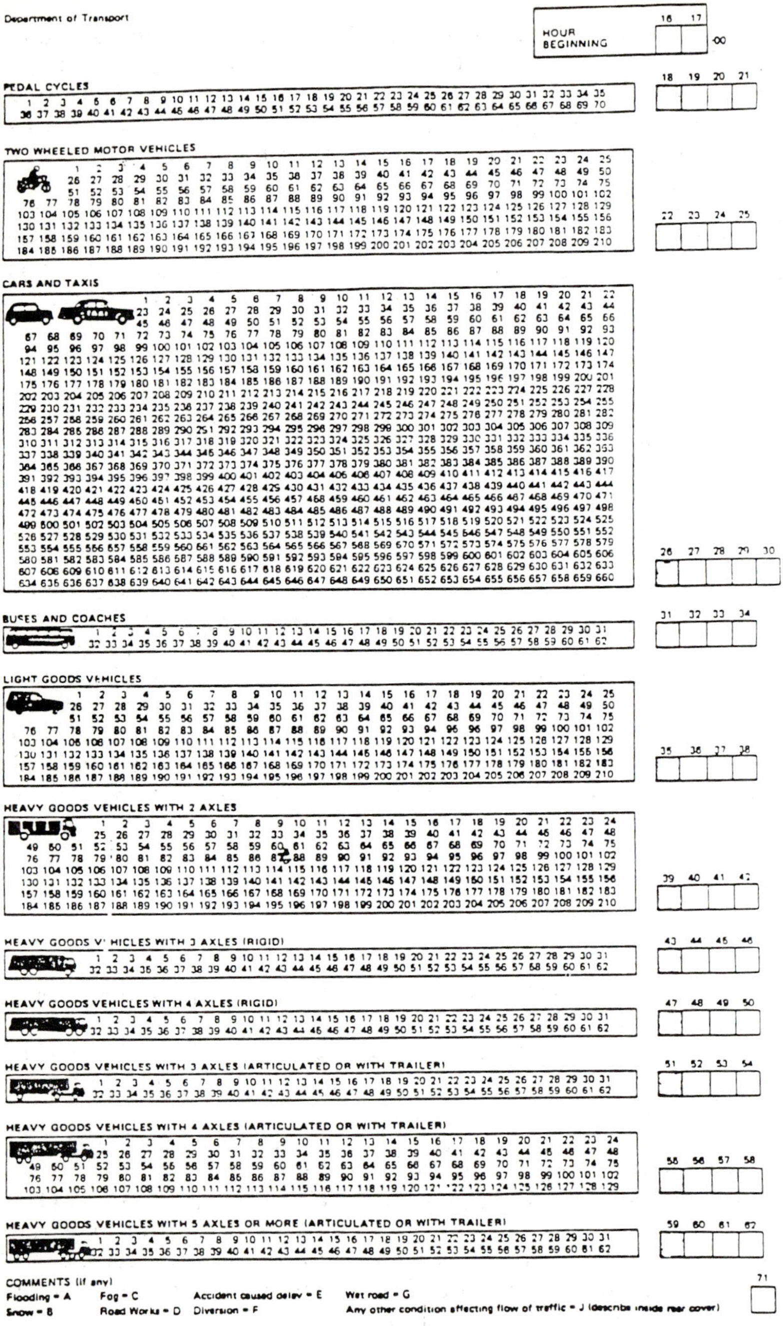

74

Department of Transport

HOUR BEGINNING | 16 17 | :00

PEDAL CYCLES

| 1 2 3 4 5 6 7 8 9 10 11 12 13 14 15 16 17 18 19 20 21 22 23 24 25 26 27 28 29 30 31 32 33 34 35 |
| 36 37 38 39 40 41 42 43 44 45 46 47 48 49 50 51 52 53 54 55 56 57 58 59 60 61 62 63 64 65 66 67 68 69 70 |

18 19 20 21

TWO WHEELED MOTOR VEHICLES

| 1 2 3 4 5 6 7 8 9 10 11 12 13 14 15 16 17 18 19 20 21 22 23 24 25 |
| 26 27 28 29 30 31 32 33 34 35 36 37 38 39 40 41 42 43 44 45 46 47 48 49 50 |
| 51 52 53 54 55 56 57 58 59 60 61 62 63 64 65 66 67 68 69 70 71 72 73 74 75 |
| 76 77 78 79 80 81 82 83 84 85 86 87 88 89 90 91 92 93 94 95 96 97 98 99 100 101 102 |
| 103 104 105 106 107 108 109 110 111 112 113 114 115 116 117 118 119 120 121 122 123 124 125 126 127 128 129 |
| 130 131 132 133 134 135 136 137 138 139 140 141 142 143 144 145 146 147 148 149 150 151 152 153 154 155 156 |
| 157 158 159 160 161 162 163 164 165 166 167 168 169 170 171 172 173 174 175 176 177 178 179 180 181 182 183 |
| 184 185 186 187 188 189 190 191 192 193 194 195 196 197 198 199 200 201 202 203 204 205 206 207 208 209 210 |

22 23 24 25

CARS AND TAXIS

| 1 2 3 4 5 6 7 8 9 10 11 12 13 14 15 16 17 18 19 20 21 22 |
| 23 24 25 26 27 28 29 30 31 32 33 34 35 36 37 38 39 40 41 42 43 44 |
| 45 46 47 48 49 50 51 52 53 54 55 56 57 58 59 60 61 62 63 64 65 66 |
| 67 68 69 70 71 72 73 74 75 76 77 78 79 80 81 82 83 84 85 86 87 88 89 90 91 92 93 |
| 94 95 96 97 98 99 100 101 102 103 104 105 106 107 108 109 110 111 112 113 114 115 116 117 118 119 120 |
| 121 122 123 124 125 126 127 128 129 130 131 132 133 134 135 136 137 138 139 140 141 142 143 144 145 146 147 |
| 148 149 150 151 152 153 154 155 156 157 158 159 160 161 162 163 164 165 166 167 168 169 170 171 172 173 174 |
| 175 176 177 178 179 180 181 182 183 184 185 186 187 188 189 190 191 192 193 194 195 196 197 198 199 200 201 |
| 202 203 204 205 206 207 208 209 210 211 212 213 214 215 216 217 218 219 220 221 222 223 224 225 226 227 228 |
| 229 230 231 232 233 234 235 236 237 238 239 240 241 242 243 244 245 246 247 248 249 250 251 252 253 254 255 |
| 256 257 258 259 260 261 262 263 264 265 266 267 268 269 270 271 272 273 274 275 276 277 278 279 280 281 282 |
| 283 284 285 286 287 288 289 290 291 292 293 294 295 296 297 298 299 300 301 302 303 304 305 306 307 308 309 |
| 310 311 312 313 314 315 316 317 318 319 320 321 322 323 324 325 326 327 328 329 330 331 332 333 334 335 336 |
| 337 338 339 340 341 342 343 344 345 346 347 348 349 350 351 352 353 354 355 356 357 358 359 360 361 362 363 |
| 364 365 366 367 368 369 370 371 372 373 374 375 376 377 378 379 380 381 382 383 384 385 386 387 388 389 390 |
| 391 392 393 394 395 396 397 398 399 400 401 402 403 404 405 406 407 408 409 410 411 412 413 414 415 416 417 |
| 418 419 420 421 422 423 424 425 426 427 428 429 430 431 432 433 434 435 436 437 438 439 440 441 442 443 444 |
| 445 446 447 448 449 450 451 452 453 454 455 456 457 458 459 460 461 462 463 464 465 466 467 468 469 470 471 |
| 472 473 474 475 476 477 478 479 480 481 482 483 484 485 486 487 488 489 490 491 492 493 494 495 496 497 498 |
| 499 500 501 502 503 504 505 506 507 508 509 510 511 512 513 514 515 516 517 518 519 520 521 522 523 524 525 |
| 526 527 528 529 530 531 532 533 534 535 536 537 538 539 540 541 542 543 544 545 546 547 548 549 550 551 552 |
| 553 554 555 556 557 558 559 560 561 562 563 564 565 566 567 568 569 570 571 572 573 574 575 576 577 578 579 |
| 580 581 582 583 584 585 586 587 588 589 590 591 592 593 594 595 596 597 598 599 600 601 602 603 604 605 606 |
| 607 608 609 610 611 612 613 614 615 616 617 618 619 620 621 622 623 624 625 626 627 628 629 630 631 632 633 |
| 634 635 636 637 638 639 640 641 642 643 644 645 646 647 648 649 650 651 652 653 654 655 656 657 658 659 660 |

26 27 28 29 30

BUSES AND COACHES

| 1 2 3 4 5 6 7 8 9 10 11 12 13 14 15 16 17 18 19 20 21 22 23 24 25 26 27 28 29 30 31 |
| 32 33 34 35 36 37 38 39 40 41 42 43 44 45 46 47 48 49 50 51 52 53 54 55 56 57 58 59 60 61 62 |

31 32 33 34

LIGHT GOODS VEHICLES

| 1 2 3 4 5 6 7 8 9 10 11 12 13 14 15 16 17 18 19 20 21 22 23 24 25 |
| 26 27 28 29 30 31 32 33 34 35 36 37 38 39 40 41 42 43 44 45 46 47 48 49 50 |
| 51 52 53 54 55 56 57 58 59 60 61 62 63 64 65 66 67 68 69 70 71 72 73 74 75 |
| 76 77 78 79 80 81 82 83 84 85 86 87 88 89 90 91 92 93 94 95 96 97 98 99 100 101 102 |
| 103 104 105 106 107 108 109 110 111 112 113 114 115 116 117 118 119 120 121 122 123 124 125 126 127 128 129 |
| 130 131 132 133 134 135 136 137 138 139 140 141 142 143 144 145 146 147 148 149 150 151 152 153 154 155 156 |
| 157 158 159 160 161 162 163 164 165 166 167 168 169 170 171 172 173 174 175 176 177 178 179 180 181 182 183 |
| 184 185 186 187 188 189 190 191 192 193 194 195 196 197 198 199 200 201 202 203 204 205 206 207 208 209 210 |

35 36 37 38

HEAVY GOODS VEHICLES WITH 2 AXLES

| 1 2 3 4 5 6 7 8 9 10 11 12 13 14 15 16 17 18 19 20 21 22 23 24 |
| 25 26 27 28 29 30 31 32 33 34 35 36 37 38 39 40 41 42 43 44 45 46 47 48 |
| 49 50 51 52 53 54 55 56 57 58 59 60 61 62 63 64 65 66 67 68 69 70 71 72 73 74 75 |
| 76 77 78 79 80 81 82 83 84 85 86 87 88 89 90 91 92 93 94 95 96 97 98 99 100 101 102 |
| 103 104 105 106 107 108 109 110 111 112 113 114 115 116 117 118 119 120 121 122 123 124 125 126 127 128 129 |
| 130 131 132 133 134 135 136 137 138 139 140 141 142 143 144 145 146 147 148 149 150 151 152 153 154 155 156 |
| 157 158 159 160 161 162 163 164 165 166 167 168 169 170 171 172 173 174 175 176 177 178 179 180 181 182 183 |
| 184 185 186 187 188 189 190 191 192 193 194 195 196 197 198 199 200 201 202 203 204 205 206 207 208 209 210 |

39 40 41 42

HEAVY GOODS VEHICLES WITH 3 AXLES (RIGID)

| 1 2 3 4 5 6 7 8 9 10 11 12 13 14 15 16 17 18 19 20 21 22 23 24 25 26 27 28 29 30 31 |
| 32 33 34 35 36 37 38 39 40 41 42 43 44 45 46 47 48 49 50 51 52 53 54 55 56 57 58 59 60 61 62 |

43 44 45 46

HEAVY GOODS VEHICLES WITH 4 AXLES (RIGID)

| 1 2 3 4 5 6 7 8 9 10 11 12 13 14 15 16 17 18 19 20 21 22 23 24 25 26 27 28 29 30 31 |
| 32 33 34 35 36 37 38 39 40 41 42 43 44 45 46 47 48 49 50 51 52 53 54 55 56 57 58 59 60 61 62 |

47 48 49 50

HEAVY GOODS VEHICLES WITH 3 AXLES (ARTICULATED OR WITH TRAILER)

| 1 2 3 4 5 6 7 8 9 10 11 12 13 14 15 16 17 18 19 20 21 22 23 24 25 26 27 28 29 30 31 |
| 32 33 34 35 36 37 38 39 40 41 42 43 44 45 46 47 48 49 50 51 52 53 54 55 56 57 58 59 60 61 62 |

51 52 53 54

HEAVY GOODS VEHICLES WITH 4 AXLES (ARTICULATED OR WITH TRAILER)

| 1 2 3 4 5 6 7 8 9 10 11 12 13 14 15 16 17 18 19 20 21 22 23 24 |
| 25 26 27 28 29 30 31 32 33 34 35 36 37 38 39 40 41 42 43 44 45 46 47 48 |
| 49 50 51 52 53 54 55 56 57 58 59 60 61 62 63 64 65 66 67 68 69 70 71 72 73 74 75 |
| 76 77 78 79 80 81 82 83 84 85 86 87 88 89 90 91 92 93 94 95 96 97 98 99 100 101 102 |
| 103 104 105 106 107 108 109 110 111 112 113 114 115 116 117 118 119 120 121 122 123 124 125 126 127 128 129 |

55 56 57 58

HEAVY GOODS VEHICLES WITH 5 AXLES OR MORE (ARTICULATED OR WITH TRAILER)

| 1 2 3 4 5 6 7 8 9 10 11 12 13 14 15 16 17 18 19 20 21 22 23 24 25 26 27 28 29 30 31 |
| 32 33 34 35 36 37 38 39 40 41 42 43 44 45 46 47 48 49 50 51 52 53 54 55 56 57 58 59 60 61 62 |

59 60 61 62

COMMENTS (if any)

Flooding = A   Fog = C   Accident caused delay = E   Wet road = G

Snow = B   Road Works = D   Diversion = F   Any other condition affecting flow of traffic = J (describe inside rear cover)

71

Fig. 6.6 Data sheet for recording classified counts (Source: Traffic Appraisal Manual)

unlikely to be many unless the survey happens to be being carried out close to a ferry terminal. Recording can be by pencil and paper, speaking into a tape recorder or hand-held data logger. If a vehicle crosses the cordon again on its way out, its journey between two cordon crossing points has been recorded. Some degree of classification can be achieved by recording the type of vehicle with a single letter, e.g. C for car, V for van. Time intervals also need to be recorded by commencing a new sheet every ten minutes, or speaking the time into the tape recorder at ten minute intervals. A data logger will probably be able to record the time of each observation from its own internal clock.

Manual recording methods are known to be fairly inaccurate with studies showing correct recording of number plates as low as 88%. Performance deteriorates in bad weather conditions.

The main disadvantage of registration number surveys is the amount of time that is taken in analysing the raw data, and this is where a computerised input device can have major advantages over manual methods. A relatively simple program can be written to match in-going and out-going numbers, but the data still has to be keyed in and the computer goes through a fairly laborious matching process.

A maximum of about four cordon crossing points is recommended for registration number surveys, otherwise the search routine that is required to trace a vehicle which has entered the survey area from its entry to its departure becomes overlong. Video recordings are able to help the surveyor by providing a permanent record of traffic movement. This allows analysis to be undertaken away from the survey site and at some later convenient time – see Section 6.18.

Prepaid postcard systems

Data can be gathered by distributing cards or forms for respondents to fill in themselves. These are then either collected directly by the surveyor, placed in a collection box or posted back to the survey team. It has been found that postal systems produce negligible returns if postage is not prepaid.

A lot of public transport surveys are undertaken in this way, with the surveyor distributing forms to passengers as they board the bus and collecting them again when they leave. The form has to be simple as the respondent will, normally, have to fill it in without assistance. Better response rates are gained on trains, where passengers can fill the form in reasonable comfort during quite a long journey, than on buses. Colour codes can be conveniently used.

Such surveys have to be carefully planned to ensure that it is feasible to gather the required data in the time available.

76

Prepaid postage questionnaires are used by the Department of Transport to gather response to public consultation exercises as part of the preparation programme for major road schemes.

Box 6.2:

A local authority wished to determine the share of income to be paid to each of several operators as part of a pre-purchase ticket scheme based on a travel card valid for all journeys by bus within the authority's area. It decided, at the same time, to gather general origin and destination data on bus travellers and the extent of use of concessionary passes for old people and schoolchildren.

Three types of concession are available:

1. Senior citizens can obtain a travel card which entitles them to half price travel at off peak times.
2. People in full time education entitled to free travel under the Education Acts can obtain a travel card which entitles them to free travel to and from school/college and half price travel at other times.
3. People in full time education not entitled to free travel under the Education Acts can obtain a travel card which entitles them to half price travel at all times.

Passengers not entitled to concessionary travel must either:

1. Hold a valid prepaid travel card, or
2. Purchase a single journey ticket from the driver.

It is decided that the survey will be carried out using a system of cards colour coded by ticket type distributed to passengers boarding the bus and collected through a collecting box as they leave. Cards have a FREEPOST address on the reverse side so that a proportion of those not collected can still be returned. All cards are numbered and handed out sequentially.

A sample of services on each route to be surveyed is determined, ranging from one service in ten on routes where there is a frequent service, to one service in three where the frequency is less than one per hour. In determining the sample, routes which are effectively variations on the same route are combined, even though they may carry a different service number. The survey is to be carried out on seven consecutive days and the sample selected to ensure that there is a good range of peak and off-peak services. Services expected to be lightly loaded will carry one surveyor, who will travel at the front of the bus close to the driver, responsible for handing out all cards and ensuring their collection. Services expected to be heavily loaded will carry a second surveyor, who will divide his/her time between assisting in handing out cards and moving round the bus assisting with completing them.

The surveyors will be equipped with three different coloured cards and a control sheet, mounted on a clipboard. The control sheet will be preprinted with each stop on the route in column 1. The other three columns will be blank and will be available for the surveyor to fill in the number of the next card of each colour each time the bus stops. The

Box 6.2 continued:

> control sheet will also include boxes for the surveyor to complete route number, date of survey, scheduled time of departure from terminus, name of surveyor, and any special information such as late running, traffic delays or bad weather conditions.
>
> The cards will be red (concessionary pass holders), green (for prepaid travel card holders) and yellow (for passengers paying the driver). Information requested on the cards will be:
>
> 1. Address of origin of the journey being surveyed.
> 2. Address of destination of the journey being surveyed.
> 3. Stop where it is intended to leave the bus.
> 4. Actual fare paid to the driver.
> 5. Journey purpose.
>
> Sections (d) and (e) will have limited alternatives and can be answered by ticking boxes.

Speed surveys

It is normal for the traffic engineer to require some measure of the mean speed of traffic over a finite distance, as he is interested in the speed at which traffic will negotiate a designed feature of the road, and he will want to know, for the purpose of estimating economic returns, the amount of delay to traffic travelling through a network. In addition, the police require measures of speed for the purpose of enforcing regulations.

The most convenient method of measuring journey speed involves the moving car observer.

A vehicle is driven several times in each direction along the section of road that is being considered. Observers in the vehicle record:

1. The number of vehicles that pass the test vehicle travelling in the opposite direction (N).
2. The number of vehicles which overtake the test vehicle travelling in the same direction (P).
3. The number of vehicles overtaken by the test vehicle travelling in the same direction (O).
4. The time taken by the test vehicle to travel through the section of road in each direction (t_w = time with the flow being measured and t_a = time against the flow being measured [hours]).
5. The time spent by the test vehicle in a queue (t_q [hours]).
6. The length of any queue (Q [km]).
7. The length of road being considered (L [km]).

Then:

Average flow in the direction opposite to that being travelled by the test car

$$q = [N + (P - O)]/(t_a + t_w) \text{ veh/hr} \qquad [6.2]$$

Average journey time of the stream

$$t = t_w - (P - O)/q \text{ hrs} \qquad [6.3]$$

Average journey speed of the stream

$$v = L/t \text{ km/hr} \qquad [6.4]$$

Average running speed

$$V_r = (L - Q)/(t - t_q) \text{ km/hr} \qquad [6.5]$$

Average queue speed

$$V_q = Q/t_q \text{ km/hr} \qquad [6.6]$$

Clearly it is important that the information gathered by this method is consistent with the use to which it is to be put. Provided reasonably steady flow conditions exist, ten runs in each direction will be sufficient to give meaningful results.

Automatic traffic counting

Automatic traffic counters can be used either to monitor traffic flows on a long term or continuous basis, or they can be installed for the purpose of a particular study.

Two systems of vehicle identification are well established: the use of an induction loop buried in the surface of the road, and a pneumatic air tube tacked down to the road surface. Others are in the course of development. Air tubes are cheap and easy to install but are subject to accidental and malicious damage. Installation of induction loops usually involves cutting and sealing a small groove in the highway surface, although it is possible to have the loops in movable mats.

An induction loop identifies the presence of a magnetic object, a vehicle, above it which operates an electronic switch in the counter housed at the side of the road. It thus counts vehicles.

With an air tube, pulses pass along the tube each time it is depressed by a load passing over it, and these are recorded by an electronic air switch in the counter. It thus counts wheels, or pairs of wheels.

In either case, the counting mechanism can be programmed to output data in various forms, or it may simply record on disk or tape which then has to be taken elsewhere for processing. A time clock can be located in the counter so that the time at which each vehicle/axle passed is recorded. Detailed advice on automatic counting is published by the Department of Transport.[4]

By careful selection of the location of combinations of loops, it is possible to identify particular vehicle classifications and turning movements, although it will normally be necessary to gather some data manually in order to determine the characteristics of the traffic stream and hence compute conversion factors.

Automatic traffic counters are generally reckoned to produce limited data to an accuracy of +/− 5%. Turner et al.[5] have reviewed the position on manual and automatic classified counting and concluded that each situation should be treated on its merits. There are no hard and fast rules for determining which approach should be used in any given circumstances.

Accident surveys

Accident statistics start from the report compiled by the police officer attending at the scene of the accident. He completes a STATS19[6] form which is passed to the highway authority to add to the accident database.

Accidents are classified according to whether the most serious injury sustained is slight, serious or fatal. Slight injuries are those which only require roadside attention. Severe injuries are those which result in the victim being admitted to hospital as an in-patient and certain other defined conditions. Fatal accidents are only those which result in the death of the victim within 30 days of the accident occurring. Some police forces also make an attempt to record damage only accidents, but since it is not legally necessary to report such accidents to the police it is always an incomplete record.

Even within the category of clearly defined injury accidents, a great deal of information is dependent on the subjective interpretation of the form by the police officer in attendance. In a serious accident, the police officer will certainly have more important things to do than fill in a STATS19 form, since his primary concern must be for the welfare of those injured and the continued safety of traffic using the road.

Whatever their shortcomings, accident statistics play a major role in determination of priorities for investment in road improvements and the economic justification of road construction.

STATS19 data includes the circumstances of the accident (weather conditions, location, time); the vehicles involved; casualties; and any special circumstances. A very simple method of recoding this data is to use large scale maps of the authority's area with clear overlays marked up to show the location of the accident, its severity and a reference number which can then be used to look up the STATS19 form or other card index system. The primary function of such a system is to identify accident 'blackspots' which can then be considered for treatment. The main disadvantage is that it is necessary to wait for accidents to happen in significant numbers before a blackspot is identified. It would obviously be preferable if the blackspot could be located before the accidents happened, so that they could then be prevented. This involves an analysis of risk, rather than simply gathering historical data.

The use of geographic information systems (GIS) can make some progress in this direction. A GIS is a computerised database in which the output is or can be related to a map or plan. Thus full details of accidents can be held in a computer file which can be brought up very quickly on a display screen. The database, like any other database, can be searched very quickly to give details of accidents having particular features or groups of features in common. Accident locations are referenced by eight figure grid references, which identifies a unique 100m square, rather than by road names or groups of road names, which can be confusing when trying to pinpoint an exact location. It should be possible, by searching the database held in a number of local authorities, to pinpoint common features in road accidents far more accurately than was possible using manual methods, and hence to make some progress towards identifying sites where there is a high risk, or a high risk could develop as a result of other changes that occur.

One such common factor that has already been identified as a cause of injury, as opposed to damage only, accidents is speed of vehicles in urban areas, and this is one of the factors that has given encouragement to engineers and government officials to press ahead with schemes to introduce traffic calming and 20mph zones. Studies have been undertaken involving observers/interviewers being on call 24 hours per day within a defined area. When an accident occurs they are notified immediately by the emergency services and attend the site of the accident. Data is therefore gathered immediately and is not the concern of the police officer, who has other primary responsibilities. Such an operation, however, is complex and expensive to mount, and it is not clear that the data gathered is of greater use than STATS19 data.

Parking surveys

Both ends of a car based trip must involve parking. With a lot of urban congestion caused by car trips, it follows that parking can be used as a method of controlling the number of car trips and, hence, the degree of congestion.

Parking facilities fall into three categories:

1. Parking on private land over which the local authority has no control other than through the application of planning consents.
2. Parking which is controlled by the local authority for the benefit of short term parkers.
3. Parking which is controlled by the local authority for the benefit of long term parkers.

Where control is exercised by the authority, two mechanisms are available for distinguishing between long and short term facilities; charges, and regulations.

In order that a local authority can develop policies for parking, it is necessary for them to have some knowledge of parking demand and provision, identified by duration, time and location. This information is necessary in order that the authority can make reasonable demands on developers in granting planning consents, and can promote sensible management policies for its own parking provision and for letting contracts to private companies. In non-metropolitan areas of England, Scotland and Wales on-street parking is entirely a matter for the county council, as highways authority, although district councils may provide off-street parking and will also act as the planning authority for development control purposes. Thus, parking is the concern of both tiers of local government. In London and the metropolitan areas of England, the Metropolitan District (London Borough) council is both planning and highway authority. Local authorities are obliged by the government to at least break even on their parking accounts. Under the Road Traffic Act 1991, management of parking enforcement in London is being transferred from the Metropolitan Police to the London Boroughs and this may well be a forerunner for changes elsewhere in the country, particularly if moves towards a unitary system of local government come to fruition. Parking can thus be seen as being relevant to a large number of public bodies, as well as commercial concerns who regard provision of adequate parking as a prerequisite of economic success.

Studies of the location of parking have to start with a large scale ordnance survey map with sufficient detail to show individual buildings and plots of land (1:1250 in urban areas or 1:2500 elsewhere). It is then necessary for an observer to either walk the area or work from aerial photographs to identify every parking space, either in use or potential. This is a major task as it will involve identifying areas such as back yards or highway verges which are or could be used for parking a vehicle, but it is not possible to make sensible decisions relating to control of parking unless there is full knowledge of all uncontrolled parking. 'Potential' spaces are important as it is likely that, when control is introduced for the first time or charges or regulations amended, a significant proportion of motorists will simply avoid the controlled area and park in the uncontrolled.

Uncontrolled parking may include:

- private car parks, whether laid out or not, belonging to factories, workshops, retail outlets and residential properties
- sections of highway to which no traffic regulation orders relevant to parking apply
- waste ground and other areas over which the authority does not exercise control.

Controlled parking may include:

- off-street car parks, which may be free, charged by pay and display, attendant or contract and may be managed by the local authority directly or by a management company acting on behalf of the local authority
- kerbside car parking, or other parking on the highway controlled by regulation (yellow lines and associated plates), disks, meters or pay and display
- private car parks, frequently associated with retail outlets, owned and managed independently but by agreement with the local authority under Section 106 of the Town and Country Planning Act 1990 (previously Section 52 of the Town and Country Planning Act 1971) entered into at the time of the planning consent and being a condition of that consent.

An estimate of the demand for parking can be obtained by adding the total number of occupied controlled spaces to the total number of occupied uncontrolled spaces and expressing this in terms of a percentage of filled capacity.

Knowledge of parking duration is important in determining the balance between long term and short term provision and in the charges that should be levied in order to maximise income and satisfy the requirement to break even on the parking account.

Average length of stay can be determined by a variation on the registration number method mentioned above. An observer starts from an identifiable point in the car park at regular time intervals and walks round the car park noting the number and year letter of the registration of each car in turn. He then returns to his starting point and repeats the process, taking care to follow exactly the same route and take approximately the same time to complete it. By comparing registration numbers from each circuit, he can work out the turnover and length of stay in each space, correct to the time interval between his visits to any particular space.

Members of the public may have a justifiable concern for the reason behind this type of activity and it is important that surveyors should carry proper identification and authorisation with them, preferably including a photocard. It is also important that proper co-ordination is maintained with the police and any private security company that may be employed on the car park.

More detailed information on origin or detailed destination of parkers will require an interview survey or the use of prepaid postal questionnaires (See above).

Pedestrian surveys

Manual counts using clipboards or tally counters, or interview surveys, are practical methods for surveying pedestrian movements. Where these are particularly concerned with crossing highways, to investigate the feasibility of a pelican crossing, for instance, it is important to be very precise about the exact location at which crossing takes place.

Freight surveys

Some information on freight movements will be available from classified counts and interviews, but these will normally be expressed in terms of lorry movements, possibly broken down by size of vehicle. More detailed information on movement of freight requires individual interviews with hauliers and operators of own-account vehicles. No specific methodology exists at present for undertaking this task.

Video surveys

The use of video cameras has a number of applications in traffic survey work.

At its most simple, a camera can be stationed close to the road and images of passing traffic recorded on tape or relayed by landline or closed-circuit television to a remote central monitoring unit. This may have a number of advantages:
1. A permanent record is made of events, which may be stored for as long as necessary.
2. Cameras can be located in positions normally inaccessible to surveyors (e.g. high-level masts).
3. Entering data for analyis can be undertaken in comfort.
4. Data analysis does not have to be in real time.

Work is in progress on developing methods of automatically sorting data from video images, particularly by Wooton Jeffries Consultants Ltd., and is reported by Hoose et al.[7]

84

Traffic data analysis and presentation

The collection of traffic data by the survey methods shown in the previous sections always leads to data storage, analysis, presentation or some combination of these three. This can involve:

1. The use of simple charts and diagrams to illustrate traffic patterns and statistics and thus make them more easily understood by technical staff, lay representatives and the general public, and to provide simple forecasting models.
2. The build up of two-dimensional matrices, primarily for use in forecasting.
3. The assignment of traffic to networks, i.e. the build up of traffic assignment models.
4. The storage of traffic data in databases and the relation of these databases to scaled plans, i.e. the use of geographic information systems.

Simple charts, diagrams and graphs

A common task for the traffic engineer is to gather together a large quantity of data and present it in a relatively simple form suitable for inclusion in a report to be considered by a client or a planning authority. The data must therefore be clear, unambiguous and accurate.

Use of pie charts, histograms and simple graphs is very common. Examples are shown in Figures 6.7 (a), (b) and (c).

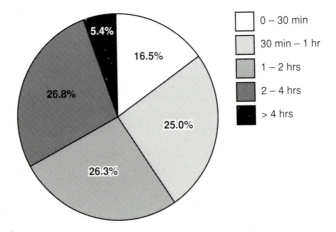

Fig. 6.7(a)   Pie chart for presentation of traffic data

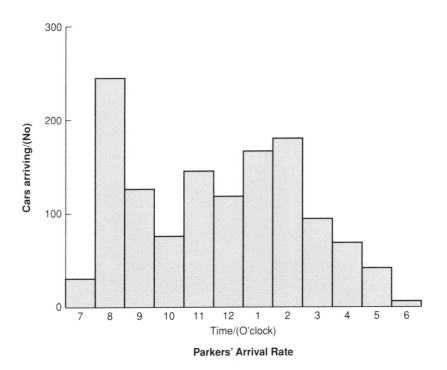

Fig. 6.7(b) Histogram for presentation of traffic data

The data can also be shown in tabular form

An alternative system of visual representation is the desire line diagram. This can simply show the origin and destination zones joined by straight lines, the thickness of the line being proportional to the traffic flow (See Figure 6.8).

Two-dimensional matrices for use in simple forecasting models

The most common form of two-dimensional matrix is an origin/destination matrix developed from, for instance, a roadside interview survey. Thus, if there are five possible origins and destinations called 01, 02, 03, 04 and 05 we might produce a simple matrix as shown in Table 6.1.

All this says is that, for instance, the number of trips surveyed with origin 03 and destination 02 was 29, so it is merely a question of summarising and presenting data in a more convenient form than

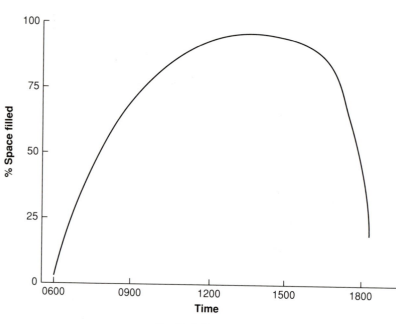

Car Park Occupancy

Fig. 6.7(c) Simple graph for presentation of traffic data

Table 6.1 Simple origin/destination trip matrix

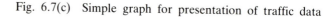

Destination		Origin				
		01	02	03	04	05
01			25	32	12	17
02		24		29	16	17
03		19		26	33	9
04		22	17	18		10
05		26	19	20	15	

would have been possible by hunting through all the original survey forms. We might have got to the same point from analysing lots of registration number records.

We can use the information by relating the number of trips to some other variable (like number of households in a zone), and hence build a trip generation model; or to relate the number of trips to a generalised cost or distance function, and hence build a trip distribution model; or relate the number of trips to a network, and hence build an assignment model.

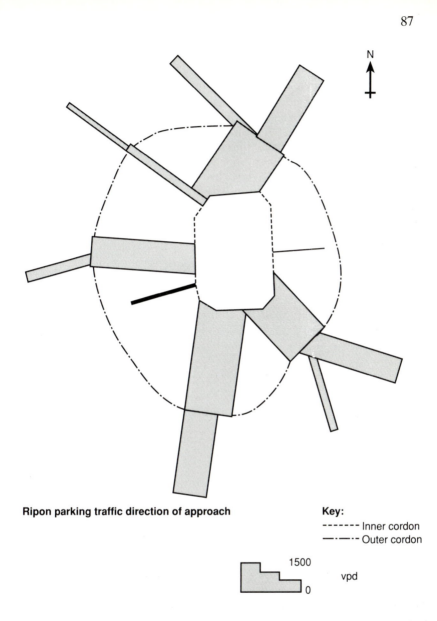

N

Ripon parking traffic direction of approach

Key:
------- Inner cordon
—·—·— Outer cordon

1500

vpd

0

Fig. 6.8 Desire line diagram

Traffic networks and assignments

As traffic engineers, we are not usually concerned with the amount of traffic that travels from A to B. We are interested in how it gets there, and the extent to which it uses the available capacity on each link in the network.

Let us consider a length of motorway, the M1, for instance, between Junction 26 (junction with the A610 north west of Nottingham) and Junction 32 (junction with M18 east of Sheffield). This section is shown diagramatically in Figure 6.9.

We have chosen a section of motorway because this is a very simple network with a very limited number of ways onto it and off it. Most sections of road do not have the same degree of control and are therefore more complicated, but the system is exactly the same.

We can simplify it a bit more by only considering the northbound carriageway and assume that nobody makes a U-turn.

Thus, starting at Junction 26, we can say that there are three ways into our network:

- from the M1 south [Stream A]
- from the A610 west (Ripley) that has turned left and come up the slip road [Stream B]
- from the A610 east (Nottingham) that has turned right and come up the same slip road [Stream C].

We can say, very simply, that traffic on the M1 before the junction with the slip road is

$$A$$

Traffic on the slip road is

$$B + C$$

Traffic on the M1 beyond its junction with the slip road is

$$A + B + C$$

If we move on to junction 27 (Junction with A608 Heanor–Mansfield), we can conclude that $A + B + C$ approaches that junction from the south because it has nowhere else to go. That traffic can either go left (A608 towards Heanor) or right (A608 towards Mansfield) or it can keep on the motorway. If we call the turning streams D and E respectively, then we know that the traffic continuing on the motorway is

$$A + B + C - (D + E)$$

We have assigned the traffic to the network.

We can make this traffic assignment process more complicated by considering alternative routes, because it is very unlikely that anyone will want to travel, for instance, from Junction 26 to Junction 32. This just represents the part of their journey that goes along the M1.

What is much more likely is that our traveller will want to go, for instance, from Nottingham to Sheffield. Depending on where he starts in Nottingham he may well join the M1 at Junction 26 and depending

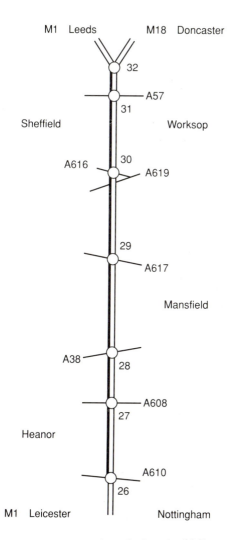

Fig. 6.9   Diagrammatic representation of a length of M1 as a traffic network

on where he wants to get to in Sheffield he may well continue on the M1 past Junction 32. We could say that, for the purpose of our model, everybody who travels from Nottingham to Sheffield will come onto the M1 at Junction 26 and still be on it past Junction 32. We would have undertaken an 'all-or-nothing' assignment, in other words, we would have identified the most straightforward (or shortest or cheapest or quickest) route and assumed that everybody did the same.

But of course some people would not behave in such a predictable fashion. They would leave the M1 at Junction 30 because they preferred the scenery or they wanted to find some cheap petrol or they were feeling tired. Our 'all-or-nothing' assignment is an approximation. If we want to be more accurate we will have to look more carefully at the alternatives and the reasons that make people select them, and develop a 'proportional' assignment. Whether this is worth doing in a particular instance depends entirely on the circumstances.

A more complex problem occurs because the same person does not always behave in the same way. Thus, driver X might go from Nottingham to Sheffield via Junction 32 today and Junction 29 tomorrow. This is because drivers will endeavour to find what they perceive to be the least troublesome route for their journey. Driver X may believe, rightly or wrongly, that the M1 north of Junction 29 becomes congested after 5.00 pm, so if his journey is earlier than that he keeps on the motorway, but if it is later he turns off. His judgement is also likely to be affected by his experience further south. If he perceives the motorway to be busy he anticipates hold-ups and turns off. On the other hand, if he perceives the motorway to be quiet, he assumes that condition will continue and stays with the motorway.

To some extent, these decisions are predictable, although the accuracy of whether the road is busy or not, when determined from a single vehicle, is likely to be very doubtful. It is not, however, the accuracy of the decision but the decision itself that determines whether an 'all-or-nothing' or a proportional assignment is a true reflection of reality.

Increasingly, we are concerned with predicting the behaviour of traffic on congested urban networks. These are considerably more complex than our simple section of the M1 and drivers are faced with a multiple choice when it comes to route selection. They will still, however, be influenced by their past experience of the routes in question and their perception of the prevailing traffic conditions. One program which seeks to predict flows on congested networks is SATURN.

Behaviour can also be influenced by signing, particularly variable message signing of the sort now being introduced to inform drivers of places where car park space is available or to alter route direction signs according to the prevailing traffic conditions.

Geographic information systems

The availability of high capacity databases and high resolution graphics monitors is making the use of GIS more readily available for storing and displaying traffic data. Networks covering substantial areas can now be brought up on screen or printed out showing flows assigned to each link. Node-based networks are more complicated as these involve modelling turning movements through junctions where time penalties might have to be applied, but they are available for small areas or individual junctions.

References and further reading – Chapter 6

1. Department of Transport. *Traffic Appraisal Manual*. Department of Transport. London. 1981. (New edition 1991).
2. Department of Transport. *Traffic Surveys by Roadside Interview*. Departmental Advice Note TA11/81. 1981.
3. Bonsall P W and Ghahri-Saremi F. *Portable data capture devices for transport sector surveys*. Traffic Engineering and Control. 28. 4: 216–223. April 1987.
4. Department of Transport. *Manual of practice on automatic traffic counting*. HMSO. London. 1981.
5. Turner D, Davies C A and Stern D. *Traffic data collection: why? when? and how?* Proceedings of the Institution of Civil Engineers: Municipal Paper 9901. 93. Mar.: 9–18. March 1992.
6. Department of Transport. STATS19 – Standard Accident Record. 1980.
7. Hoose N, Vicencio M A and Zhang X. *Incident detection in urbanroads using computer image processing*. Traffic Engineering and Control. 33. 4: 236–244. April 1992.

Chapter 7

Traffic Forecasting

Uncertainty

Determination of the size of future traffic flows is an essential input to the design of transport facilities. Clearly, it would be ridiculous to provide a single track road with passing places in a situation where traffic volumes would be likely to fill up a three lane motorway. It simply would not work: traffic would grind to a halt, find alternative routes, or not make the journey. Equally, providing a three lane motorway where half a dozen cars and a push bike might come along would be hopelessly uneconomic and unacceptably damaging to the environment.

So a means has to be devised to determine how much traffic is likely to use a particular link on a transport network in order that the physical nature of that link can be determined.

This is not a problem unique to highway designers, although highways perform a greater number of functions than other links on the network and forecasting future use is therefore more complicated. However, British Rail must forecast likely use of a line before investing in improvements; airlines must forecast likely passenger levels before buying new aeroplanes; airport authorities need to know the number and type of aircraft that will be using their terminals before they can raise the necessary capital and set their architects and engineers to work to build them.

Forecasting is not something that should come as a surprise to civil engineers. Most of the time, we are concerned with estimating the loads that will be applied to a structure, including its own weight, applying a suitable factor of safety and then designing the structure to combine materials of known properties in such a way as to ensure that loads will be carried safely. In most civil engineering structures, however, the major part of the load to be carried is constant through their life. There are certain variables, like wind loading or snow loading, and some structures have to cope with things like tidal conditions, but overall it is economic for the structure to be designed to cope with the predictable worst case. Failures are very rare, and when they occur, they tend to be catastrophic, certainly in economic terms and sometimes in terms of injury and loss of life.

In highways, we are concerned with the traffic 'load', although it is measured in veh/day rather than kN/m^2. It is rarely economic to design a highway to cope with the maximum load to which it is likely to be subjected (although the same may not be said about other transport facilities).

Throughout the life of a highway, the load, or the amount of traffic using it, will be continually varying. Traffic levels vary by time of day, day of the week and time of year. Over time, they tend to increase, so if we plot a graph of total traffic in the United Kingdom, measured in terms of vehicle kilometres, it will be of the shape shown in Figure 7.1. We can, and do, gather statistical data on which we can base graphs

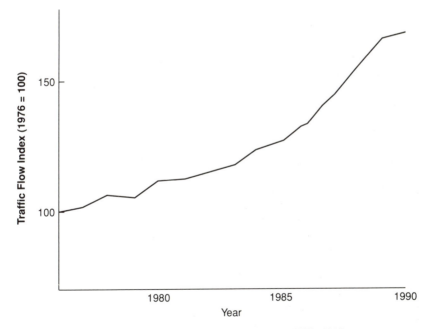

Fig. 7.1 UK traffic growth in vehicle kilometres, 1970–1990

like that shown in Figure 7.1. Up to a point, this is a statement of fact (within the limitations of statistical accuracy). Everything that has happened up to the present moment is a statement of fact. There are certain limitations involved in gathering, analysing and presenting the data, so Figure 7.1 only represents the facts as they were up to and including 1989, but there is an essential difference between fact and forecast, and that difference happens *now*.

As soon as we move into the future, by years, or days, or even seconds, we enter into the realms of forecasts and there must, therefore, be some measure of uncertainty. A scientific approach to forecasting must seek to minimise this uncertainty. It should not seek to become a substitute for fact. It is very important that a highway engineer who, by his training, is used to dealing with facts, should appreciate this point. Otherwise, he may not be able to appreciate the economic, environmental and political aspects of design.

A look at the sad tale described in Box 7.1 shows that any traffic forecast must reflect *uncertainty*. There is no way that we can be absolutely certain about anything that is going to happen in the future.

Box 7.1:

The best laid plans . . . a case history

An engineer employed by a local highway authority was appearing as a witness into a proposal to construct a new dual carriageway section of 'A' road. The proposal to construct the road as a dual carriageway had been taken following a full COBA appraisal using the National Road Traffic Forecasts. Unfortunately, in order to maintain consistent standards along the length of the road, it was necessary to pass through the corner of a marshy area known to be the breeding ground of a particularly rare species of wildlife, and designated as a Site of Special Scientific Interest.

Somewhat further along the route, it passed close to the boundary of a new estate of four-bedroomed detached houses, the occupiers of which were very concerned at future noise levels from a high speed road relatively close to their properties. Although they would be eligible for double glazing to be installed under the terms of the Land Compensation Act, this would not be much help on hot summer days when they wanted the windows open and to go sunbathing in the garden. What they wanted was the road reduced to a single carriageway, thereby reducing speeds, and a noise barrier, suitably landscaped, inserted on the ground where the second carriageway would have been. They therefore founded the Oak Close Protection Society and, being relatively prosperous, hired a barrister on their behalf to appear at the public inquiry.

The barrister fully appreciated the importance of consistent standards along the length of the road, and felt that the strongest case could not necessarily be made by pleading for the sunbathing rights of the middle classes, particularly since the road proposal was popular among many residents of terraced houses who would have heavy lorries removed from

Box 7.1 continued:

their front doorsteps. He therefore homed in on the SSSI and its about-to-be-dispossessed newts, correctly establishing that a single carriageway alignment could be built without encroaching on the marshy area. At no point did he mention that it would also meet all his clients' objections, although he did establish that there would be considerable financial savings ('enough to pay for reducing the environmental impact of the scheme'); it would be popular with the local nature groups and would not offend the terraced householders.

The question revolved around the validity of the traffic forecasts. The Engineer had carried out extensive surveys and, using COBA perfectly correctly, had built up a trip matrix for the design year 2010. This had produced the result that the forecast traffic (correctly expressed as 24 hour Annual Average Daily Traffic) was 19,124 vehicles per day, and, in accordance with the recommendations contained in Department of Transport Standard TD20/85, a design using a dual two-lane carriageway was prepared, assessed and found to give a satisfactory rate of return.

In cross-examination, the barrister concentrated on the figure of 19,124. He established that it was the correct figure, there were no typing errors, and, yes, that was the figure that the computer had produced for the forecast traffic figures in 2010.

The barrister then spent an hour and a half reducing the engineer almost to tears, in the way that barristers can do. The line of the argument was that 19,124 was indeed between 19,123 and 19,126, and that it actually represented a range of values which didn't make much difference to the principle of the design. Until, that is, the barrister had managed to widen the range to something between 17,999 and about 20,000. At that point, the barrister, who had done his homework, produced a copy of TD20/85 and pointed out, correctly, that a traffic flow of up to 18,000 vehicles per day should be used as input for a single carriageway assessment.

The mistake that the engineer had made was failing to appreciate the nature of the traffic forecast and trying to present it as a fact. The consequences were very serious. As a result of the inquiry, the road was down-graded to a wide single carriageway, newts and sunbathers were saved but the authority is already having to undertake emergency (albeit low cost) works in an effort to reduce the accident rate.

This is not to say that we can not make rational decisions concerning the future. We can use logic, together with historical knowledge, to reduce the degree of uncertainty that is involved in any particular forecast.

National road traffic forecasts (NRTF)

We might wish to consider the global level of all traffic in Great Britain, measured in terms of vehicle–kilometres. This is a very general measure of traffic levels. It is an important statistic and is used by government to determine, for instance, the amount of income they can expect from motor taxation and petrol taxes. Relative levels of increase, derived from NRTF, are used to predict future traffic flows on particular lengths of road if no more locally derived data are available.

All forcasts must be based on historical data. We are not in the business of reading the stars or consulting oracles. We are concerned with measuring historical facts, determing the relationships between those facts, projecting those relationships into the future and hence determining our forecasts.

It has been found, historically, that global traffic levels depend largely upon two variables: the gross domestic product (GDP) of the country and the real price to the user of petrol. Projecting such forecasts into the future implies that the 'desirability' of owning a car (and other consumer goods which give rise to increases in goods traffic) remains more or less constant. This amounts to an assumption that people will choose to purchase a car as soon as they can afford to do so, and that the amount they use it will depend, to a large extent, on the price they have to pay at the pumps for petrol. Real disposable income of individuals does vary with GDP.

There are, of course, limitations on the growth of car ownership. There is a proportion of the population who are too old, too young or who suffer infirmities which make car ownership an unlikely or unachievable ambition. Motor manufacturers operate in a free market and will not perceive it as being desirable to achieve 100% saturation. They seek to maximise profitability, not sales volume. They may well concentrate on refining exisiting models, thereby making them more expensive and ensuring that they remain out of reach to a proportion of the population. There will always be some people who are too poor to buy a car, and whilst GDP, and hence, the disposable income of the poor, may increase, so will the sophistication, and the price, of a motor car.

This gives rise to the idea of a 'saturation' level of car ownership, i.e. the level at which all those who want, and can afford, a car actually have one. Tanner[1] developed such an approach at the Transport and Road Research Laboratory, assuming saturation levels of 40%, 50% and 60% and combining them with growth in GDP of between 66% and 108% and increases in the real price of petrol of between 54% and 200% to produce a series of curves showing increases in traffic volume between 1976 and 2000. These curves are shown in Figure 7.2.

The method developed by Tanner is basically that which is used by the Department of Transport today. The figures are updated approximately once a year, and published in the form of the National Road Traffic Forecasts. These break down traffic volumes by category of vehicle and show high and low forecasts in the form of indices up to

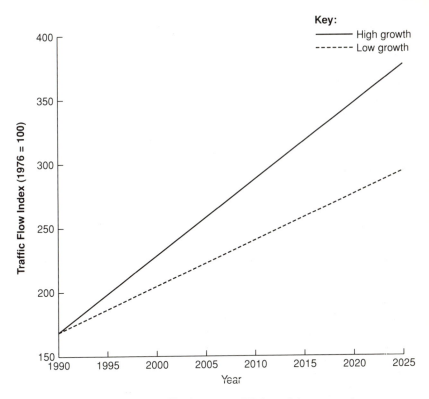

Fig. 7.2 National Road Traffic forecasts. High and low growth

the year 2025. An extract from the most recent NRTF tables is shown in Tables 7.1 and 7.2.[2]

These tables are very simple to use. In order to determine the future level of traffic on any particular section of road, it is first necessary to determine the present traffic flow. This can be done by conducting a traffic survey. If the road in question does not yet exist (e.g. in the design of a proposed by-pass) it is not a difficult matter to run a computer program that will tell us how much traffic would use the road if it did exist: this is a version of traffic assignment and is a well established procedure in transport planning.

In the interests of standardisation, traffic planners use particular measures of traffic flow. One of these is the Annual Average Daily Traffic (AADT), which is used to determine capacity of rural roads, and, hence, the standards to which they should be designed.

AADT is defined as being the total amount of traffic which uses a section of road in a year divided by 365. It can refer to all vehicles, or to particular categories of vehicle, e.g. cars or light goods vehicles. Definitions of various types of vehicle, and in the classification into which they fall, are contained in the COBA Manual, but for our

Table 7.1 National Road Traffic Forecasts – Low Forecasts

Year	Car	LGV	OGV(1)	OGV(2)	PSV	Total
1976*	1.000	1.000	1.000	1.000	1.000	1.000
1989*	1.717	1.611	1.124	2.152	1.385	1.671
1995	1.948	1.814	1.179	2.449	1.385	1.887
2000	2.126	1.991	1.229	2.744	1.385	2.054
2010	2.481	2.397	1.335	3.423	1.385	2.398
2025	3.001	3.168	1.511	4.709	1.385	2.929

* = Observed values

Table 7.2 National Road Traffic Forecasts – High Forecasts

Year	Car	LGV	OGV(1)	OGV(2)	PSV	Total
1976*	1.000	1.000	1.000	1.000	1.000	1.000
1989*	1.717	1.611	1.124	2.152	1.385	1.671
1995	2.104	1.952	1.219	2.677	1.385	2.030
2000	2.421	2.276	1.316	3.217	1.385	2.334
2010	3.016	3.097	1.533	4.586	1.385	2.933
2025	3.806	4.913	1.926	7.613	1.385	3.827

* = Observed values

purposes it is only necessary to give a general description of the two Other Goods Vehicle categories, OGV(1) and OGV(2). Broadly speaking, these are vehicles larger than a local delivery van. Of this group OGV(2)s are the largest type of heavy goods vehicle. The indices contained in the table assume a significant transfer of goods from small to large vehicles, consequent upon the gradual increase of maximum permitted vehicle weights to bring the UK into line with the rest of Europe.

In order to determine AADT, it is not necessary to start a survey at one minute past midnight on January 1 and continue until midnight on December 31. AADT is not a measured value; it is a statistic, a value derived from measurements and then adjusted by the application of other measurements. The exact method of derivation of AADT is laid down by the Department of Transport and published in the Traffic Appraisal Manual.[3] Traffic surveys must be conducted on particular days (usually a Tuesday or Thursday) in a neutral month (March, April, May, September or October). Counts can be undertaken for 12 hours (0700 to 1900) or 16 hours (1600 to 2200) and are then factored up to provide a value for total annual traffic. The factors used depend on the month of the survey and the type of road, and are based on the results of a large number of surveys conducted over the past 40 years or so by the Transport and Road Research Laboratory.

E Factors, which are used to translate 12-hour flows into 16-hour flows, and M Factors, used to translate 16-hour flows into Annual flows, are shown in Tables 7.3 and 7.4 respectively.

Once the present AADT value for the road is established, it is a simple matter to project it forward to any year in the future using the NRTF tables. For instance, if it is desired to work out a forecast value of AADT for all traffic in 2010 based on a value of 20,000 vpd (vehicles per day) in 1989 based on low growth, referring to Table 7.1 we can see:

1989 index, all vehicles = 1.671
2010 index, all vehicles = 2.398
$$AADT^{2010} = (AADT^{1989} \times Index^{2010})/(Index^{1989})$$

Therefore $AADT^{2010} = (20000 \times 2.398)/1.671$
$$= 28701$$

Table 7.3 E Factors (Source COBA9 Manual p. 4–27. November 1986)

Road Type	E Factor
Main urban	1.18
Inter-urban	1.15
Recreational/inter-urban	1.13

Application:
If a road is classified as main urban and a 12-hour count has established flow in both directions as 12,873 vpd, then statistic to be used for 16-hour flow

$$= 12873 \times 1.18$$
$$= 15190$$

Table 7.4 M Factors. (Source COBA9 Manual p. 4–29. November 1986)

	Main urban	Inter-urban	Recreational/Inter-urban
April	371	407	464
May	361	391	416
June	369	385	371
September	367	371	351
October	365	390	417

Application:
A road is classified as recreational/inter-urban and a 16-hour count undertaken in May has established a flow of 14507 vpd. What is the AADT?

Annual flow = 14507×416
$$= 6.035 \times 10^6$$

AADT $= (6.035 \times 10^6)/365$
$$= 16534$$

There is another advantage of this approach to traffic forecasting. By concentrating on a level of saturation determined by the maximum level of car ownership, it effectively reduces the importance of time, and hence of the rate at which traffic will grow. The little calculation that we have carried out above develops a forecast for the year 2010, but the time component of this calculation is purely nominal. We have chosen the year 2010, partly because it is a nice round figure and it appears in our reduced table of NRTF indices, but primarily because it is a likely choice of 'design year' for a road being designed in the early 1990s. This is because recommendations on the 'size' of a length of rural road are based on AADT in the fifteenth year after opening.[4] A road being designed in, say 1992, might be expected to go out to contract in 1993, take two years to build and open in 1995. The fiifteenth year after opening is therefore 2010.

The implications of this calculation are two-fold. It is primarily used as a means of determining the government's priorities in allocating funds to building roads. It is used to calculate the COBA rate of return (see Chapter 25), i.e. as one input to the decision to construct Scheme A, Scheme B or Scheme C. It is also used in the design of the road, but the important point here is that the road is designed to carry, approximately, 28,000 vpd; not that it is designed for 2010. The fact that the flow rate of 28,000 is expected to be reached in 2010 is incidental. The authors of NRTF may have been over optimistic in forecasting the growth rate of the country: maybe we will have a depressed economy for the next ten years, and traffic growth is slower than expected. In which case, the 28,000 vpd will not be reached until 2015. But the assumption implicit in the Department of Transport's traffic forecasting procedures is that it will be reached eventually.

The forecasting of future levels of road traffic is a highly controversial subject. There are a wide variety of political arguments which surround it, ranging from those who would have us believe that traffic will continue to grow indefinitely and that we must invest the maximum possible amount of resources to building roads, to those who believe that the only reason that traffic grows is because we build roads, and that if we didn't build any new roads at all, traffic would cease to grow. Neither of these extreme positions is likely to find much favour with engineers who are concerned with finding feasible solutions to measurable problems.

The history of the past twenty years has been that actual traffic growth, in global terms, has tended to be nearer the high forecast than the low one. Petrol prices have tended to fall steadily in real terms since the so-called oil crisis of the early 1970s when the oil producing countries, co-ordinating their action through OPEC, managed to force a sudden increase in the price of oil to the developed countries. Despite the panic that was caused at the time the OPEC countries were unable to sustain their position, partly because there were always those of their number who were prepared to break ranks in order to capture a larger share of the market and partly because of the interdependence of the developed and developing countries of the world. As far as Britain is concerned, we became an oil producing country during the same period as North Sea oil reserves came on stream.

The price of petrol to the consumer is largely determined by government action, as the greatest proportion of the price paid is tax. So far, no British Government has had the courage to introduce a large general increase in petrol taxation in the interests of reducing fossil fuel consumption and vehicle emissions which are environmentally damaging. This may change with increased public awareness of environmental issues and with directives emanating from the European Commission. However, dramatic change is unlikely: draconian increases in petrol tax would be highly inflationary, and with the British economy universally regarded as being very weak, the last thing that we can afford is a return to high levels of inflation. The freedom of action of any British Government is therefore very limited. The most that we can expect in the short to medium term (the 20 years or so over which we are concerned to forecast traffic growth) is petrol prices which either stay level or increase slowly, with governments concentrating on less radical measures such as tax incentives to use smaller cars or a move away from flat rate taxation (Vehicle Excise Duty). This, in turn, suggests that traffic growth in line with the high forecasts is probably not very far from what is going to happen unless very positive action is taken to restrain it.

Whilst there is clearly a relationship between the price of petrol and a decision to use a car for a particular journey, it is by no means a question of direct proportionality. A draconian increase in petrol prices would not lead to a sudden reduction in the amount of traffic on the road as we all left our cars at home. The decision is made at the point of purchase of the car. Whilst some public transport operators have tried to wean drivers from their cars, and some local authorities have tried to make the use of the car prohibitively expensive for certain journeys, the overwhelming choice of travellers is to use a car for any journey for which a car is available. This problem is aggravated where the car concerned is a company car, and the marginal cost to the user of using it is either nil or very low.

Other individual choices have a major effect on travel patterns. A feature of the past two or three decades has been a move in the population from inner city areas to out of town suburbs and country villages. It is not unusual to find people commuting sixty miles and more on a daily basis to get to work. They do this, at considerable cost to themselves, in order to gain what they perceive as the advantages of living space or fresh air or education for their children or whatever. Unfortunately, areas of low population density (out in the country, for want of a better expression) are very difficult to serve by public (communal) transport. Thus, the increasing proportion of the population who do not live in city centres is increasingly dependent on private cars. The price of petrol could rise to £10 per gallon and they would still be so dependent. They might cut out the odd trip to take the children to the seaside, but that would not significantly reduce the global level of traffic on Britain's roads.

The effect of overall economic performance

We have stated, above, that the major inputs to NRTF tables, are the price of petrol to the user and the GDP – the general state of the economy of the country. In times of boom, travel levels go up; in times of recession, they stay the same or come down. Actual global traffic volumes began to fall in the third quarter of 1990 after rising steadily for several years. This coincided with what some economic indicators described as the beginning of the worst recession to hit this country since the Second World War. If government could see its way to producing a steady and sustained recession in the economy, we would also see a fall in the level of traffic on the roads!

Of course, no government sets out to produce recession. The language of democratic politicians is the language of growth, of rising living standards and, inevitably, more traffic on the roads.

The conclusion that follows is that traffic forecasts must take on board the certainty that global traffic levels will continue to grow because they are based on fundamental tenets of life in an advanced democracy. There may be periodic hiccups, such as that in 1972–73 (caused by OPEC action) and 1990–92 (caused by economic mismanagement) but these will eventually work themselves out of the system, because the alternative is that the system itself will break down.

So the forecasts suggest that we are heading for ever increasing demands on our overcrowded road system, and there is good reason to suppose that the forecasts might be right, at least in general terms.

Political alternatives

Politicians have reacted to this situation in different ways, depending on their political credentials. The Conservative Government has massively expanded the road construction programme, primarily by giving priority to schemes which are relatively simple to implement. This has involved widening on existing routes and construction of by-passes, neither of which is likely to run into serious public opposition. Mindful of the inflationary effect of large scale public expenditure, they have sought to introduce private capital to the road system. They have also sought to bring about an expansion of public transport, although not all would agree with their methods, through bus deregulation and privatisation of municipal operations and their proposals for reorganisation of British Rail. They have toyed with the most likely mechanism for traffic restraint in urban areas, road pricing, and encouraged traffic calming. There have been developments in London such as Red Routes and the appointment of a London Traffic Supremo responsible to the Secretary of State. Rapid transit proposals in Manchester, London and Sheffield have progressed.

Much of this work is threatened by general economic recession and pressure on the PSBR. Grant aid under Section 56 of the Transport Act 1968 for the West Midlands Rapid Transit system has been postponed and the Jubilee Line extension into East London is threatened by the collapse of Canary Wharf developer, Olympia and York, who intended to contribute £400 million to the cost of linking their prestigious tower to the tube network.

Perhaps unsurprisingly, the Labour Party has poured scorn on the Conservatives' ideas. They tend to place a greater emphasis than the Conservatives on the role of public transport and promise a high level of investment in the railways, a shift of funding away from roads and an end to privatisation. However, despite the rhetoric, there does appear to be a high level of consensus amongst politicians of all parties. Whoever exercises power, it is likely that we will see investment continue to be poured into inter-city transport facilities by both road and rail, whilst the cities are likely to see developments of public transport facilities combined with measures of traffic restraint.

Local forecasts

Whilst no aspect of transport will ever be entirely divorced from politics, we are normally concerned with a more immediate problem: how much traffic is going to use the particular bit of the transport network that we are currently designing?

We have seen above how NRTF tables can be used to predict traffic levels on particular sections of road, starting off from survey data or computer model output. The choice between high and low forecast levels is a political choice and reflects the politicians' view on how much money there is available and how important are roads compared to other calls on the public purse. As engineers, we may have reservations about the politicians' decision, but it is not a technical problem. As the case study in Box 7.1 indicates, we should beware of being tied down to treating forecasts as if they were facts. In any case, as we have again seen above, the difference between high and low forecasts does not affect the ultimate traffic level, only the speed at which we arrive there.

The procedure of basing traffic forecasts on survey data and NRTF indices is valid in some circumstances. In order to determine the AADT (Annual Average Daily Traffic) 15 years after opening a new section of rural road in order to identify a category of road as required by TD20/85, this procedure may well be the one to use provided that we do not have any information available which suggests otherwise.

Local highway authorities now have very sophisticated databases covering traffic flows on all main links in their area. These can be updated from the results of a rolling programme of traffic surveys and implementation of major land use changes. Existing systems are largely road based and do not include forecasts of modal split which will be necessary if public transport expands and is able to attract people out of their cars.

Box 7.2:

Calculating a traffic forecast

A major commercial development is proposed with access off an existing local authority primary distributor road. The new development, which will be primarily offices, will have 1000 car parking spaces provided for workers. The developer has agreed in principle to enter into an agreement under Section 106 of the Town and Country Planning Act 1990 to pay for the upgrading of the existing local authority road for a length of approximately 1 km from the site boundary, on the grounds that the existing road is not adequate to cope with generated traffic and the value of the developer's site will increase with improved access facilities. The existing two-way traffic flow on the local authority road is 750 veh/hr in the peak hour. To what standards should we improve the primary distributor?

There are two parts to this problem. We need to predict the 'natural' growth in traffic on the primary distributor, and then add the generated traffic which is due to the development.

We need to make one or two assumptions. Assume the design year is 2010 (15 years after completion of the new road).

The term 'primary distributor' does not tell us a great deal. It is usually used in the context of a hierarchy of urban roads, but very few urban hierarchies conform to a theoretical ideal, so we can take the term as being a very general description of the road's function. Crucially, we will assume that it is not subject to a speed limit of 50 mph or less: it therefore comes under our definition of a rural road.

That being the case, it is reasonable to suppose that NRTF values apply. In this question, we have not been told anything different. Of course, there may be three other developments along this length of

Box 7.2:

road. That would be different (and our developer would probably be less than keen to pay for his rivals' improvements). There might be a coal mine along there somewhere, scheduled for closure in two years time. That would be different – but we don't know any different, so we assume that NRTF figures apply.

We are looking at traffic flows in the peak hour, not at AADT. This is probably a reasonable thing to do, assuming that our office workers go to work in the peak and go home in the peak. But it is an issue the developer is likely to want to raise with the local authority since it has implications for the cost he is going to have to bear.

Doing the same calculation we did in Section 7.2 (and assuming the 750 veh/hr refers to 1989) but using high growth figures we can work out that, by 2010, 750 veh/hr will have grown to 1316 veh/hr.

We then have to add in the generated traffic. We will have to make some decisions about how full the car park is going to be and what the distribution of arrivals and departures is going to be within the peak hour. One thousand vehicles in an hour is not the same thing as 16.67 vehicles every minute or one vehicle every 3.60 seconds. But for the purpose of this illustration, let us assume that value of 1000 veh/hr is a realistic representation of the generated traffic.

The total traffic forecast is therefore:

$$1316 + 1000 \text{ veh/hr} = 2316 \text{ veh/hr}.$$

Urban congestion

It is unlikely that forecasts based on NRTF tables will be applicable in urban areas. Changes in land use patterns are far more likely in towns and cities in the short term, and the network itself is likely to be more complex. Thus, changes to one part of the network are likely to have far reaching consequences. A situation like that envisaged in Box 7.2, where a single development can be seen as being the only major change to a single link 1 km long is unlikely to occur in an urban area.

It is now generally accepted that there is no way that towns can be adapted, in all but the most unusual circumstances, to cope with all the traffic that wishes to use the roads in an urban centre. This realisation has taken a long time to gain acceptance – the nature of the problem was originally described in the Buchanan Report[5] in the 1960s – but the capacity of an urban road network is now accepted as being a major, perhaps the major, limitation on future traffic growth.

Put very simply, the demand for travel may well grow in a manner similar to that described in Figure 7.1 towards a saturation level of 50% car ownership. But it cannot transform into road traffic, which is what we are trying to forecast, if it cannot be physically fitted on to the roads that are provided for it because they are already full. As engineers, we may wish to conduct our forecast by means of a computer model, by which we seek to assign traffic to particular links on a network. This is a Traffic Assignment Model, and it is a relatively simple procedure up to the point where the capacity of a particular link is reached. It then becomes meaningless to assign any more traffic, because it will not fit. We might say that the capacity of Link A–B is 200 veh/hr. Our survey tells us that the present flow is 175 veh/hr. We can undertake a projection based on NRTF tables and establish that, by 2010, the flow will have grown to 307 veh/hr. Carrying our analogy with rural roads forward, this would mean that we would have to provide additional roadspace – widen the road – to cope with the additional demand. But in most urban situations, this is not practicable: land is too expensive, buildings must not be knocked down and likely public opposition would render the scheme impossible to construct.

So we are stuck with a capacity of 200 veh/hr. Our forecast must become 200 veh/hr. But what happens to the remainder of the demand?

Experience in London has shown that drivers are amazingly tolerant of congestion. It would make life much simpler if, as soon as the network approached congestion, drivers left their cars at home and went by bus. But they do not, and although efforts are now under way to try to make communal transport more attractive to potential car drivers and passengers, these are still at a relatively experimental stage and it remains to be seen what measure of success can be attributed to them once they are fully operational.

Faced with an intolerable level of congestion, however that might be defined, the most likely thing for a driver to do is to change his travel patterns, i.e. to make the same journey but at a different time. Congestion is usually at its worst during the morning and evening peaks, i.e. when commuters are travelling to work and going home again. Most people start work between 8.30 and 9.00 in the morning and finish between 4.45 and 5.15 in the afternoon. In a small town where journeys to work are short and capacity is only reached at a few points on the network and for short periods, these times are likely to define the peaks.

Figure 1.3 shows how traffic distribution through the day varies with the size of the town being considered. As the size of a conurbation becomes larger, delays become longer and more drivers are forced to change their journey patterns in order to avoid intolerable delay due to congestion. This results in 'spreading' the peak. In a major city, we may well find that traffic flows continue at capacity through the peak hour and beyond.

In London the situation is unique, with traffic building up very early in the morning, and peak flows being maintained throughout the day, not falling off again until late evening.

We are only just beginning to develop assignment models which are able to deal with this aspect of driver behaviour. Only when this has been successfully achieved will it possible to add on the basis of modal choices. In the meantime, developments such as Manchester Metro have to be based on commercial judgements, which implies some degree of risk.

So the problems of forecasting future traffic levels are essentially different, depending on whether the road is rural or urban. Capacity restraint is usually crucial in an urban situation. In rural areas, the intention at present is to expand the amount of road space available to meet forecast demand.

That demand is likely to go on increasing, although the rate of increase will largely depend on the performance of the national economy.

References and further reading — Chapter 7

1. Tanner J C *Car ownership trends and forecasts*. TRRL Report LR799. Transport and Road Research Laboratory. Crowthorne. 1977.
2. Department of Transport. *National Road Traffic Forecasts*. DTp. London. 1989.
3. Department of Transport. *Traffic Appraisal Manual*. Revised edition. London. August 1991.
4. Department of Transport/Welsh Office. Departmental Standard TD20/85. *Traffic flows and carriageway width assessments*. Department of Transport. London 1985.
5. Buchanan C (Chair). *Traffic in Towns*. HMSO. London. 1967.

Chapter 8

Route Planning and Selection

The origins of a major highway scheme

A new road can have its origins in a variety of places. If it is part of the national road network, i.e. it is going to be a trunk road, then it will be up to the Government and the appropriate Department (Transport, Wales, Scotland, Northern Ireland or, possibly, Environment) to initiate the scheme. In very broad terms, it will have to be in accordance with the policy of the Government of the day, which, at present, is to expand the inter-urban road network as quickly as funds can be made available and the necessary legal procedures can be completed.

Certain types of scheme are afforded priority, including the provision of by-passes to small towns and those which contribute to growth in development areas.

In preparing its roads programme, the Government has to be mindful of various pressures, particularly those which come from Members of Parliament, local authorities, pressure groups, the media and the general public. In England, the Department of Transport has a number of regional offices, part of whose duties is to keep a finger on the pulse of informed opinion.

Every so often the Government publishes its views on the needs for road construction in the form of a White Paper[1] which allows the possibility of a parliamentary debate. The decision to include a particular scheme in the roads programme is a matter for Parliament.

Once that decision has been made, the need for the scheme is deemed to have been established, although a lot of detailed decisions about its design and its route still have to be made. The inclusion of a particular scheme at this stage is no guarantee that it will be built in the year for which it is initially proposed, or, indeed, in any year. The records of strategic transport planning are full of outline schemes long since consigned to the wastepaper basket.

Apart from major schemes proposed by central government, there are a host of other ways in which a project can come into being. On all roads that are not trunk roads, the highway authority is the District Council in Metropolitan areas of England, the county council elsewhere in England and in Wales and the Regional Council in Scotland. Each highway authority will have a rolling programme of roads schemes which will have come about through the procedures required for Structure Plans and Unitary Development Plans, through the annual Transport Policies and Programmes submission to central government or through *ad hoc* arrangements within the authority to identify need.

Additionally, developers may take the initiative in new road construction if it is required to provide adequate access to the development or if other considerations arise in the negotiations between the developer and the planning authority. Procedures for local authority roads are, within reason, similar to those which have to be followed on a national basis.

The development of a preferred scheme

The current procedure for bringing a road proposal from initial inclusion in the government's programme to the point where work can start on site is as follows (this table is based on that contained in the 1986 SACTRA Report[2]).

1. Identification of need. It is necessary to identify the ways in which the scheme will satisfy the Government's objectives, which may be to reduce journey times, reduce accidents, reduce congestion and/or improve safety.
2. Scheme identification study, which must:
   - Define the area that will be affected by the scheme by means of a map.
   - List the problems that will be solved by the scheme. Where possible, these should be quantified.
   - List the national objectives that will be served by the scheme.
   - List the local objectives that will be served by the scheme.
   - Report the views of the public, if these are known.

- List the alternative options that are available for meeting the objectives and explain how they will be evaluated.
- Include an outline of economic assessment.
- List any further data that will be required for the evaluation to be completed satisfactorily.

3. Entry into the Roads Programme in the form of a White Paper. This is not a firm commitment, merely a decision to proceed with further studies.

4. Appointment of consulting engineer or agent authority. The Department of Transport has now decided that it will only make an appointment in respect of initial studies. Appointments for detailed design and site supervision will be made separately and later. Where an agent authority is appointed, this reflects the expertise available within the authority and implies that the authority will, to all intents and purposes, perform the same role as a consultant might elsewhere. The Government has been keen to move away from traditional forms of contract for road construction, which it believes discriminates against low-cost bids. It has therefore sought to introduce design and build into highways contracts, in which a contractor bids for the whole of the work, from this stage onwards.

5. Initial surveys and consultation with the Landscape Advisory Committee, local authorities and other statutory bodies. The first stage of the traffic, environmental and economic assessment is carried out using COBA9[3] and the Manual of Environmental Appraisal[4] (MEA).

6. Local model validation. Forecasts of future traffic flows are prepared. There is an inital COBA9 run, which will involve determining the future trip matrix and and a 'first shot' at a COBA rate of return.

7. Framework drawn up in accordance with the recommendations of MEA.

8. Technical appraisal draws together all options to be put to public consultation.

9. Public consultation: this provides an opportunity for members of the public to view alternative options at an exhibition, to discuss their implications with members of the design team and to take part in a questionnaire-based survey on which option should be preferred.

10. Preliminary report to include the technical appraisal and the results of the public consultation, make a recommendation of the preferred option, stating reasons.

11. The Secretary of State announces the preferred route. This action protects the line of the proposed road from alternative development and authorises the start of detailed design work by the Department's consultants or agent authority.

12. Detailed surveys, design and assessment are carried out. At this

stage, environmental protection work must be identified and included in costings. All previous work is updated.

13. COBA output is reviewed along with local model validation procedures.
14. The framework is updated.
15. The state of the Line and Side Roads Orders is checked and reported to make sure all procedures are complete.
16. Internal approval of the scheme for inclusion in the Works Programme.
17. Draft Line Order and Side Roads Order is published together with details of how, where and to whom objections may be made.
18. Review of COBA and Framework, together with all other evidence, in preparation for a public inquiry. An Inquiry will be necessary if objections have been received from statutory objectors or persons affected by the works to the draft orders and not withdrawn following negotiation.
19. The Public Inquiry, held before an Inspector appointed by the Secretaries of State from a list of suitable persons maintained by the Lord Chancellor.
20. The Report of the Inspector to the Secretaries of State for Transport and the Environment.
21. The Secretaries of State decide to confirm, amend or reject the draft orders and publish their decision.
22. If the decision of the Secretaries of State has involved significant amendments to the scheme then it will be necessary to go back through the above procedure to the point where the amendment is no longer significant.
23. Compulsory Purchase Orders are drawn up and published, if this has not already been done under the earlier procedure. This may lead to a further public inquiry.
24. Once all the Orders are confirmed, it is only necessary to get final approval and financial allocation to proceed with the work.

The above stages are those followed by the Department of Transport when preparing proposals to improve trunk roads or build new sections of the inter-urban trunk road network. Local highway authorities are expected to go through a similar process, although fixing the line of a new highway will require planning consent under the Town and Country Planning Act 1990 rather than the making of an order under the Highways Act. The calling of Public Inquiries and confirmation of Orders is still a matter for the Secretary of State.

The general procedure is one of starting off with a very broad brush approach, in which all possible solutions are considered, and then gradually whittling it down until the preferred solution is identified.

The criteria used for rejecting certain options may involve anticipated problems with land acquisition, environmental impacts or constructional difficulty. A scheme will not be considered unless it generates a certain minimum rate of return as measured by COBA.

Public consultation

A particular issue in the planning process for new roads is the requirement for public consultation. The basic procedures are set out in a Department of Transport Leaflet.[5] The need for the scheme is, by this stage, deemed to have been established. Public consultation only concerns alternative routes, a matter of some frustration to political lobbyists who are opposed to the whole of the Government's Roads programme. Where there is no alternative route, i.e. only one of the routes originally devised in the development programme is deemed to be feasible, a public consultation exercise may still be mounted in order to compare the feasible scheme with a 'do minimum' alternative, i.e the effect on the existing network if no new road is built. In the case of the M1–M62 link in West Yorkshire, which is included in the Roads Programme, the public consultation exercise has generated such an amount of well co-ordinated hostility that the Department of Transport is understood to be considering adopting the 'do minimum' alternative.

The public consultation stage revolves around three means of communication between the Department and the public: the consultation document, the exhibition and the questionnaire.

The consultation document is a glossy, full-colour leaflet which can be obtained free at the exhibition, local council offices, public libraries and sometimes post offices. It is distributed to anyone seen by the Department to have a legitimate interest, including pro- and anti-road pressure groups. It will give very general information on costs, the broad environmental effects and other relevant factors.

The exhibition consists of a series of displays, usually mobile, which are set up at predefined locations on particular days. Suitable venues are village halls, hotels, etc. The intention is that anyone who lives or has a business close to the line of a proposed new road should have an opportunity to visit the exhibition and consider the proposals.

The questionnaire is supplied with the consultation document, and members of the public are invited to express their views on the alternatives and return the questionnaire either before they leave the exhibition or through the post. Postage is prepaid.

Public inquiry

After the public consultation stage, the Department will consider all the evidence at its disposal and decide upon the preferred route. This will be announced, giving an opportunity of public discussion, but there is no further formal procedure unless and until a public inquiry is called.

It is a matter for the Secretary of State for Transport and the Secretary of State for the Environment to jointly consider whether it is necessary to hold an inquiry. If there have been objections to the compulsory purchase of land which have not been withdrawn, there must be an inquiry. It is unlikely that a major road scheme would get through all the necessary procedures without an inquiry, but it might be possible to do without for minor schemes, thereby speeding up the preparation time and minimising preparation costs.

An inquiry is held before an Inspector, who is appointed by the Secretaries of State on the recommendation of the Lord Chancellor. The Inspector is independent, and he is not an employee of the Department of Transport. He may be a lawyer, an engineer, a planner, an architect or a member of some other relevant profession. His job is:

1. to conduct the inquiry observing principles of openness, fairness and impartiality;
2. to report on the material points of objection presented to him;
3. to find and record the relevant facts;
4. to present his views on the merits of the arguments for and against the proposals;
5. to recommend whether or not the proposals should be approved with or without modification or give his reasons for not making any modifications.[6]

The conduct of the inquiry is largely at the discretion of the inspector. People who are directly affected by the proposals have a right to be heard, as do representatives of local authorities and statutory undertakers. In practice, the Inspector will allow anybody with a valid point to make to be heard, although he may disallow repetitive evidence or obstructive behaviour. The Inspector will not permit questions of government policy to be raised at a local inquiry, and this includes the often contentious matter of national road traffic forecasts.

The proposing authority – usually the Department of Transport or a county or district council – will speak first and outline the case for the proposal. They may call witnesses in support of their argument, must answer questions put to them by the Inspector and be available for questioning by objectors or their legal representative. Other statutory bodies will then be invited to put their case, either for or against the proposals, followed by organised groups and finally, individuals. Every participant is entitled to legal and/or professional representation, and is subject to questioning by the Inspector and other participants. At a major inquiry, the proposing authority is likely to be represented by a barrister specialising in planning law, so cross-examination can be a harrowing experience. Inspectors, however, are there to see fair play and it is not unknown for the Inspector to come to the rescue of an engineer being subjected to unreasonable legal pressure. Finally, the proposing authority will sum up and have the last word.

Inspectors may be accompanied by one or more assessors, who are specialists in particular factors being considered at the inquiry.

It is important to note that the Inspector does not make the decision on the outcome of the inquiry. He is there to report and to recommend, but the decision is taken by the Secretaries of State and there is no obligation on them to accept the recommendations of the inspector. For major schemes, this will mean the Secretary of State in person. Minor schemes may effectively be determined by a junior minister, but the final decision is political and it is the politicians who should take ultimate responsibility for the location and design of a new road.

At the point when line orders are published, if not before, property affected by the proposal will become subject to planning blight, and, in appropriate circumstances, owners can serve notice on the proposing authority under the provisions of the Town and Country Planning Act 1990 requiring the authority to purchase the property. Planning blight is not only expensive to the proposing authority if a large number of properties are involved: it can also lead to the general rundown of the area, closure of businesses and dereliction of property. It is therefore in everybody's interests that the planning procedures for new roads should be completed as quickly as possible.

Various other factors may be relevant to the public perception of a new or improved road, and may therefore be relevant to the inquiry. Various rules apply under the Land Compensation Act 1973 for compensating people whose homes, farms and businesses are adversely affected by new roads where the construction of the road would involve purchase of land or property and where the authority has powers to compulsorily purchase.[7,8,9]

In addition, highway authorities are required to provide insulation in private residential property under the terms of the Noise Insulation Regulations.[10]

Insulation by means of double glazing, double doors, ventilation and venetian blinds must be offered to occupiers of property within 300 m of a new road or road with a new carriageway added where

1. the traffic noise level at one or more facades will increase by at least 1 dB(A) and will not be less than 68dB(A) L10(18 hour) and
2. noise caused or expected to be caused by traffic using the new or altered section of road will contribute at least 1 dB(A) to the noise level.

Any claims must refer to traffic levels no more than 15 years after the opening to traffic of the new road or new carriageway and they can only refer to new roads or carriageways opened after 17 October 1972. Highway authorities have discretionary powers in respect of road improvements which do not constitute a new road or new carriageway, and in respect of roads opened between 17 October 1969 and 16 October 1972.[11,12]

Private sector initiatives

It has been the intention of Conservative Governments elected in 1979, 1983, 1987 and 1992 to encourage private participation in the development of road and other inter-urban transport projects. The reason behind this thinking is twofold:

1. One of the most universally acclaimed successes of the Thatcher premiership was the privatisation of several major industries, including steel, gas and telecommunications. It may not be possible to entirely privatise the road network but it is possible to privatise the financing, construction, maintenance and operation of clearly defined major routes with limited access.
2. By encouraging the use of private, rather than public capital, it reduces demand on the Public Sector Borrowing Requirement and makes for a healthier economy.

The Government's intentions were set out in a White Paper[13] published in May 1989. This emphasised the Government's belief in the value of measures '. . . to harness the entrepreneurial, financial and management skills of the private sector to the provision of roads.' The private sector, however, has been rather less keen than the Conservative Government to see the wholesale introduction of private sector capital. The amounts involved are regarded as too great and the return on investment through tolls seen as too risky to be tackled during a time of recession. There are, however, one or two exceptions.

The Queen Elizabeth II Bridge, which carries the M25 London orbital motorway across the Thames at Dartford, was built on the basis of a design, build and partial finance contract by Trafalgar House in return for a concession to levy tolls on both the bridge and the adjacent tunnels for 25 years.

The second Severn crossing, a cable stayed bridge to be built at English Stones, downstream of the existing suspension bridge, will also be built on the basis of a single design and build contract, and the contractor will recoup his outlay by charging tolls. A similar arrangement is being negotiated for the Birmingham Northern Relief Road. The powers to facilitate these arrangements were taken by the Secretary of State under the terms of the New Roads and Street Works Act 1991.

A further example of attempts to introduce private capital into transport facilities is provided by the Channel Tunnel and the associated link to London. The tunnel itself was the subject of special legislation enacted for the purpose, which allowed the floating of a company, in which several million people have invested their savings. The London link, now seen as the scheme which will eventually be constructed, started off as a proposal from consulting engineer Ove Arup. The legislation which established the structure for the construction and operation of the tunnel and the link to London specifically precludes the Governmenent from investing public funds.

The extension of the Jubilee Line on the London Underground was approved on the basis of a £400 million contribution by Olympia and York, the Canadian company developing Canary Wharf, which would be served by the new extension. The failure of the developer in 1992, before the complex was complete or construction of the underground line had been started, has placed a question mark over the viability of the Jubilee Line project.

References and further reading – Chapter 8

1. Roads for prosperity: the Government's roads programme. White Paper. Cm. 693. HMSO. London. May 1989.
2. Williams T E H (Chairman). *Urban Road Appraisal – The 2nd Report of the Standing Cimmittee on Trunk Road Assessment* ('SACTRA'). HMSO. 1986.
3. Department of Transport. COBA9 Manual. 1981 (updated).
4. Department of Transport. *The Manual of Environmental Appraisal.* Department of Transport – APM Division. 1983. (Under revision as at May 1992).
5. Department of Transport, *Trunk Road Planning and the Public – the procedures outlined.* Department of Transport. London. 1991.
6. Department of Transport. *Public Inquiries into Road Proposals – what you need to know.* Department of Transport. London. 1990.
7. Department of the Environment. *Land Compensation – Your Rights Explained. Booklet 2.* Your home and nuisance from public development. HMSO. 1990.
8. Department of the Environment. *Land Compensation – Your rights explained. Booklet 3.* Your business and public development. HMSO. 1991.
9. Department of the Environment. *Land compensation – your rights explained. Booklet 4.* The farmer and public development. HMSO. 1991.
10. Noise Insulation Regulations 1975. SI1975/1763 as amended by Noise Insulation (Amendment) Regulations 1988. SI1988/2000.
11. Department of the Environment. *Land compensation – your rights*

explained. Booklet 5. Insulation against traffic noise: the noise insulation regulations 1975. HMSO. 1991.

12. Department of Transport. *Calculation of Road Traffic Noise.* HMSO. 1988.

13. New roads by new means – bringing in private finance. White Paper. Cm 698. HMSO. 1989.

PART III

LAYOUT DESIGN

Chapter 9

Highway Link Design

The road between junctions

The term 'Highway Link Design' is taken from the title of the Department of Transport Standard, TD9/81,[1] and its accompanying Advice Note, TA43/84.[2] It comes from the theoretical treatment of a series of highways as a network, which is made up of nodes and links. The link is therefore the section of road between nodes, rather than that between junctions, although for most purposes, nodes and junctions are the same thing. The only difference may arise when a particular road is insignificant in traffic terms, and will therefore not show up on the traffic network. Figure 9.1 illustrates nodes and junctions on a highway network.

The function of a civil engineer in relation to the design of a highway is to specify its geometry and the materials of which it is made in such a way as to ensure that it can be used safely and conveniently by members of the public. He has the job of preparing drawings and tabulations that will define, precisely and in three dimensions, the location of each and every part of the structure of the road. Specification of materials is outside the scope of this book. A recent text covering British practice on that subject has been provided by Watson.[3]

From the point of view of the designer, a highway can be considered to be made up of two types of component:

(a) The road between junctions;
(b) The junctions themselves.

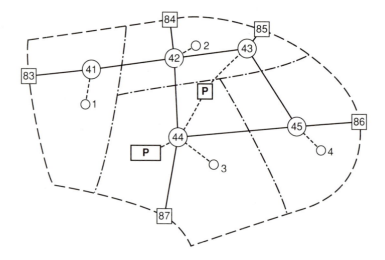

Fig. 9.1 Nodes and links on a highway network. **Key:** —— cordon, —·—— zone boundary, ⟋○³ zone centroid, ⑭ node, ⟦87⟧ cordon crossing point, ⟦P⟧ car park

Between junctions, the designer will be concerned with a number of activities which might be taking place. In a simple situation, vehicles will be travelling along the road at a reasonably constant and predictable speed. This will happen at many rural locations on the main road network, where parking on the carriageway may well be prohibited; there are few, if any, pedestrians; accesses will be limited and junctions will be widely spaced and carefully designed. There will be few constraints on the range of design solutions which can be selected by the engineer, and he will primarily be concerned to work out a horizontal and vertical alignment that can be used safely by traffic travelling at high speed. The quantity of traffic that will use the road will be relevant, for a decision will have to be made between, for instance, single and dual carriageways, but this will follow from traffic forecasts and scheme appraisal, and will not form a major part of the geometric design. The design team will be more concerned to balance the amount of cut and fill involved in the earthworks, and locate any major bridges or retaining walls in the best positions.

In other situations, particularly in urban areas, the designer's job can be very different and, in some respects, more complicated. An urban street has many uses. These are described in some detail in

Chapter 10, but for now it is sufficient to consider that traffic on an urban road may well travel at variable speed and there will be frequent conflicts with parked vehicles, pedestrians and cyclists. Junctions, and the resulting intersections between opposing or merging traffic streams, will be frequent and are often unplanned and unco-ordinated. The designer must seek to satisfy conflicting demands for space, and his freedom of action may be very limited. Large scale earthworks are unlikely to form part of an urban highway scheme, and while some structural work may be inevitable, its significance in costing is likely to diminish when compared with other costs such as land, demolition and service diversions.

British practice, then, divides highways into two categories for the purpose of adopting design standards: rural roads and urban roads. The difference between them is defined as being that an urban road is a road on which a speed limit of 50 mph or less is imposed. Otherwise, it is a rural road.

The traditional approach to defining the geometry of the highway involves preparing a plan at a scale of 1:500 or 1:1250 of the area through which it is to pass. Then, working at a drawing board, the designer, armed with railway curves, transition tables and templates, can draw a proposed horizontal alignment onto the plan. A number of alternatives might be considered at this stage, particularly if the project is to go through a public consultation exercise. The designer can use his knowledge and experience to judge whether a particular alignment is likely to be satisfactory. Constraints, such as areas of bad ground or suitable locations for bridges, will be known to the designer and he can take them into account. The alignment is usually defined by its centre line, so considerable care has to be taken where it passes through cuttings or over embankments as the total amount of land taken up by the road depends on the steepness of the side slopes. Separate alignments may be designed for each carriageway of a dual carriageway if cost savings result.

Having fixed the horizontal alignment, or possibly a series of alternative alignments, existing ground levels are required along the proposed centre line in order that the designer can fix the vertical alignment so as to meet the required standards for sight distances, gradients and vertical curve lengths while approximately balancing the amount of cut and fill. Any survey undertaken for highway purposes must have sufficient detail to allow contours to be drawn at intervals of no more than 0.5 m across the line of the road or to allow precise ground levels to be interpolated wherever the centre line is eventually fixed. Earthworks form a major item in the cost of a new rural road scheme and its viability is therefore likely to be affected by the relationship between existing and finished ground levels.

Finally, cross sections can be drawn at, say, 25 m intervals along the alignment so that channel, footway and verge levels can be fixed, allowing for appropriate drainage falls and superelevation. Drawings showing existing and proposed plans, a longitudinal section and cross sections are required as part of any contract to construct a section of road.

Using modern computerised technology the procedure is very similar, although the designer uses an interactive graphics screen and mouse, or other input device, in place of a drawing board. Most major road schemes now utilise the program MOSS which is more fully described in Chapter 14. The advantage of utilising Computer Aided Design (CAD) is that it speeds up the many repetitive calculations that are inherent in the design of highway geometry. MOSS contains facilities to develop alignments using continuous transition curves (cubic splines) where the mathematics involved is too complex to be undertaken manually. Strictly speaking, it is not necessary to develop cross sections in MOSS, as the program is able to calculate accurately volumes between two surfaces using isopachytes. The program can, however, produce sets of cross sections exactly similar to those produced on a drawing board, and this is often felt to be a useful facility to help the designer visualise the geometry he is creating.

MOSS is generally regarded as being an essential tool for the design of major, rural road schemes in Britain, and many of the features of UK Government standards are built into the program.

There are a number of other programs which are derivatives of survey drawing packages and some mention of these is made in Chapter 14. They are particularly suitable for determining the layout of estate roads, and can be used to add features such as roads, footways, car parks, buildings and planting areas to drawings produced by the surveyor. Designers therefore have a choice between a very powerful package – MOSS – with almost unlimited memory capacity which requires a relatively expensive workstation installation, and a number of lower cost packages which can be run on PCs but which may not have the capacity to handle the quantity of data involved in a major highway project. The degree to which CAD is used in highway design depends largely on the availability of hardware and software in a particular office.

Nearly every local highway authority in the UK is known to have I-MOSS (Interactive MOSS), usually mounted on a graphics workstation or workstations. The availability of staff willing and able to use it is, however, less certain.

Estate road layouts are often designed by architects or by developers' in-house design teams. In either case, capital costs involved in adopting CAD can mitigate against its installation. Architects rarely see 'externals' (i.e. roads, car parks, landscaping) as being much more than a necessary add-on. Developers are concerned with cash flow and since, by the nature of the job, income comes at the end, they are unlikely to be very enthusiastic about expenditure at the beginning. It may be unfortunate that highway engineers are often not involved with this type of job until something has gone wrong.

Rural design standards

The procedure for establishing the alignment of a new road can commence many years before construction starts. The length of this 'lead-time' has been a matter of concern to successive governments, but efforts to reduce it have not been entirely successful.

In order to promote a new road, the highway authority must obtain planning permission and purchase any land that is necessary for the construction of the road. If the road is a trunk road, the designers must also go through the public consultation procedures that are outlined in Chapter 8. Once all these hurdles are successfully passed, it is necessary to obtain funding. In an effort to minimise the effect of some of these obstacles, the Government has resorted in recent years to a number of tactics, including concentrating on widening existing roads rather than building new ones, advancing schemes which are politically popular, like by-passes to small and historic towns, investigating the feasibility of privately funded roads and placing special legislation before Parliament (the Channel Tunnel).

Many of the decisions which the designer must make are made in the context of scheme appraisal, rather than as part of a recognisable design process. This includes the selection of the standards that are appropriate to a particular length of road. It is clearly helpful for the designer, and all other people and bodies concerned with decision making, to be able to work within a range of standard layouts. This simplifies the appraisal process and makes computer application possible. Available standards for rural links are laid down in Department of Transport Standard TD20/85,[4] Traffic Flows and Carriageway Width Assessment. Table 2 of the standard is reproduced as Table 9.1.

There might be a number of stages during the preparation of a scheme at which the designer has to prepare an outline design, or possibly a number of alternative designs, in order that the appraisal process can be undertaken. This will certainly be necessary at public consultation stage and at any public inquiry, as a 'COBA rate of return' (see Chapter 25) will have to be calculated and a framework drawn up and published. The main justification for road construction in the UK is based on the ratio of costs and benefits of a particular scheme. It may well be that a particular design becomes a preferred option, not because the benefits it is perceived to generate are greater than the alternatives, but because the ratio of benefits to costs gives the best rate of return. Thus, appraisal is a very important part of the preparation process and may well preclude some solutions which a designer would otherwise wish to incorporate.

The left-hand column of Table 9.1 can be seen to list the seven alternative carriageway layouts from which designers are able to choose, they are:

Table 9.1 Flow levels for rural link assessment

Road class (COBA classification) (i)	24 hour AADT flow 15th year after opening (ii)	Edge treatment (iii)	Access treatment (iv)	Junction options relating to flow		
				Minor road junction (v)	Major road junction (vi)	
Normal single carriageway S2 (1)	Up to 13,000*	1 m hardstrips	Restriction of access. Turning movements concentrated. Clearway at top of the flow range	Simple Junctions or Ghost Islands (Ref 9)	Ghost Islands, single lane dualling or roundabouts (Ref 8, 9, 10 & 11)	
Wide single carriageway WS2 (2)	10,000 to 18,000	as 1	as 1	Ghost Islands, single lane dualling or roundabouts (Ref 8, 9, 10 & 11)	as 1	
Dual 2-lane all purpose carriageways D2AP (3)	11,000 to 30,000	as 1	as 1	Priority Junctions. No other gaps in the central reserve	Generally at-grade roundabouts	
	30,000 to 46,000**	as 1	Restriction of access severely enforced and left turns only. Clearway	No gaps in the central reserve	Generally grade separation	

Road class (COBA classification) (i)	24 hour AADT flow 15th year after opening (ii)	Edge treatment (iii)	Access treatment (iv)	Junction options relating to flow	
				Minor road junction (v)	Major road junction (vi)
Dual 3-lane all purpose carriageways D3AP (3)	40,000 and above	as 1	Restriction of access severely enforced and left turns only. Clearway	No gaps in the central reserve	Generally grade separation
Dual 2-lane Motorway, D2M (4)	28,000 to 54,000**	Motorway standards	Motorway regulations	None	Grade separation
Dual 3-lane Motorway D3M (5)	50,000 to 79,000**	as 4	as 4	as 4	as 4
Dual 4-lane Motorway D4M (5)	77,000 and above	as 4	as 4	as 4	as 4

* Upper limit of flow range assumes maximum diverting flow of about 2,000 vpd during maintenance works.
** Upper limit of flow range assumes maximum diverting flow of about 10,000 vpd during maintenance work. (See para. 4.2.)

Normal single carriageway, S2
Wide single carriageway, WS2
Dual 2-lane all purpose carriageway, D2AP
Dual 3-lane all purpose carriageway, D3AP
Dual 2-lane motorway, D2M
Dual 3-lane motorway, D3M
Dual 4-lane motorway, D4M

The dimensions of each of these carriageway layouts is shown in another Department of Transport publication, the Standard Highway Details.[5] Sketches showing the main dimensions are shown at Figures 9.2 (a) to (g) and photographs of the more common standards are shown at Figures 9.3 (a) to (d).

Single two-way carriageways will normally be 7.3 m wide + 2.0 m edging strips, although many variations are possible. In particular, very lowly trafficked routes in remote areas may only justify a single track road with passing places.

Wide single carriageways are contentious and some engineers will not entertain them on safety grounds. It is recommended that they should only be considered in circumstances where visibility is extremely good, or on very short links where the additional width may aid manoeuvrability. Note that this standard refers to a single carriageway, usually 10 m wide between kerbs or edging strips, divided into two lanes. There is no scope within the standard for a 3–lane carriageway, and where these have been inherited they are in the process of being changed to have two-lane markings. A distinction should be drawn between a WS2 carriageway (10 m wide divided by lane markings down the centre) and a S2 carriageway + crawler lane on an uphill gradient (divided into three lanes but having clear directional priority).

The distinction between motorways and all purpose roads is an administrative one, in that a motorway is a road to which a Special Roads Order made under the Highways Acts applies. The main features of a motorway are that there is no frontage access, there are substantial distances (about 5+ km) between junctions, there is a hard shoulder for most of its length and all cross traffic is grade separated. Certain types of vehicle are prohibited from using a motorway.

Selection of standards for heavily trafficked roads is rarely left to the designer. It is far more likely to be a political decision, heavily influenced by economics. Thus, the M40 extension from Oxford northwards past Banbury to Birmingham is a dual 2-lane motorway (D2M) even though predicted traffic volumes suggested each carriageway should have three lanes. At the time the vital decisions were being made, public expenditure was being severely curtailed, so the money for the third lane was not available. A graphic example of the same reasoning can be seen on the A1 at Wetherby, West Yorkshire. During 1988–89, a diversion involving a new bridge over the River

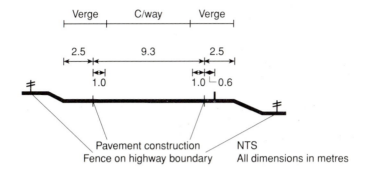

Fig. 9.2(a)   Cross-sectional details: S2 carriageway

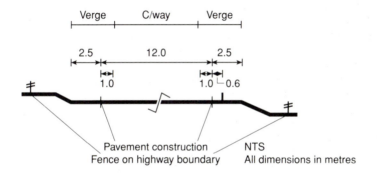

Fig. 9.2(b)   Cross-sectional details: WS carriageway

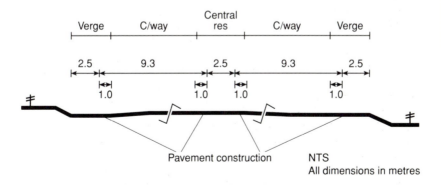

Fig. 9.2(c)   Cross-sectional details: D2AP carriageways

130

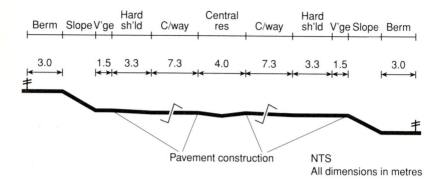

Fig. 9.2(d)   Cross-sectional details: D2M carriageways

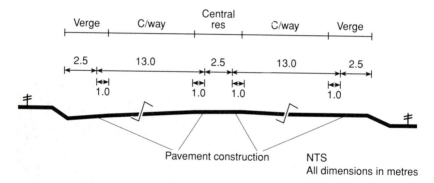

Fig. 9.2(e)   Cross-sectional details: D3AP carriageways

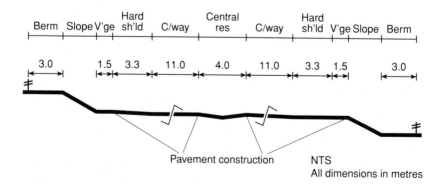

Fig. 9.2(f)   Cross-sectional details: D3M carriageways

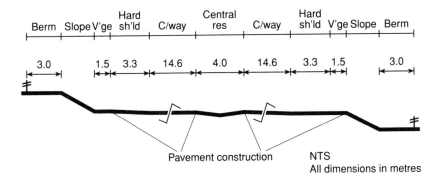

Fig. 9.2(g) Cross-sectional details: D4M carriageways

Fig. 9.3(a) S2 carriageway

Wharfe was constructed in order to avoid a roundabout constructed on the original Wetherby By-Pass in the 1960s. The diversion is a dual two-lane all purpose road (D2AP). In 1990–91, the section of A1 immediately south of Wetherby is being reconstructed as part of the government-inspired proposals to upgrade the A1 to motorway standards throughout its length. This section will have three lanes.

It seems inevitable that the main trunk road system will continue to defy consistent planning and continue to reflect the priorities, and the economics, of whatever government happens to be in power at the time of detailed planning and design.

132

Fig. 9.3(b)   S2 carriageway with crawler lane

Fig. 9.3(c)   D3M carriageway

On roads of less strategic importance, however, the engineer may well find himself with an opportunity to influence decisions on the

Fig. 9.3(d)   WS carriageway

overall standards adopted for particular highway links. In this situation, the starting point for determining layout standards is the forecast traffic flow measured as 24-hour AADT (Annual Average Daily Traffic) in the 15th year after opening. A definition of AADT and the method of its determination is given in Chapter 7. The significance of the 15th year follows from the discounting technique used in COBA, by which the present value of benefits diminishes the further into the future that the benefits accrue. It should be recognised that uncertainty increases with time into the future and that traffic forecasts therefore become less and less reliable. Fifteen years is seen as being a reasonable time period over which to appraise new road construction, even though certain aspects of design, such as the pavement structure, may look further ahead.

The second column of Table 9.1 shows the range of flows for which a particular standard of layout might generally be considered appropriate. Other features of the layout (edge treatment, access treatment, minor and major junctions), are shown in columns (iii) to (vi). Thus, having located the forecast traffic flow in column (ii), the road width can be determined from column (i) and the other features from the remaining columns.

It is important to recognise that this is not a design guide: it is a starting point from which alternative designs can be developed and fed into the appraisal process.

A study of the table, however, will show the logical result that, as traffic flows increase, the standards of the road built to carry it will also increase. So too, probably, will the construction and maintenance costs.

Within any given design standard there is still a great deal of opportunity to vary particular features, and hence to improve the cost effectiveness, as measured by the COBA rate of return, of the design. Junctions and accesses, in particular, can be designed in such a way as to minimise construction costs and maximise benefits. It is important that the designer appreciates the way in which his actions can influence the appraisal of the project as a whole.

Box 9.1:

Here is an example of the way in which political and economic factors can affect design standards:

A section of trunk road was built between 1987 and 1989 in order to provide a by-pass to a small industrial town in the north of England. The town, and its industry, had developed in the bottom of a steeply-sided valley, so the by-pass had to be located on high ground with steep gradients at either end. Heavy earthworks were involved in its construction, including one substantial rock cutting.

Estimated traffic flow (24 hour AADT) in 2004 was 24,000 vpd, including 4,500 Heavy Goods Vehicles (19%).

On the basis of Table 2, TD20/85, this represented a fairly clear case for a D2AP road with restrictions on access, priority junctions and roundabouts at either end of the by-pass.

There are no properties having or requiring access but in other respects the road, as built, is almost unrecognisable in terms of Table 2. It is a S2 carriageway, although on the gradients, which constitute about 60% of the length, a crawler lane has been added giving a carriageway width of about 10 m. Two minor roads cross the line of the by-pass and grade separation has been provided for both, with no access between the minor road and the by-pass. At one end of the by-pass there is a priority intersection with turning lane, acceleration and deceleration lanes provided on the main road. At the other end, there is grade separation with all turning movements catered for by slip roads.

The explanation is entirely concerned with money. At the time that decisions were being made regarding standards to be adopted, government policy was to restrict public expenditure as tightly as possible. Very few trunk road schemes were constructed at that time. Because of the difficult ground over which the by-pass had to be constructed, adding an extra carriageway would enormously increase the cost. By restricting the width of the road, bridges could be single, relatively short spans, and advantage could be taken of level differences in the natural ground to provide grade separation at relatively low cost. Because the by-pass is a relatively simple link on the network offering clear advantages over the congested High Street route, there was not much doubt that the vast majority of by-passable traffic would (and

Box 9.1 continued:

does) use the by-pass. So improved standards would not significantly increase the amount of traffic using the by-pass and benefits, as measured by COBA, are largely proportional to the amount of traffic using the new route. Benefits, in any case, are limited by the steep gradients, so time savings are not as great, at least for HGVs, as they would be on flatter terrain. The difference in speed, and hence journey times, between light vehicles and HGVs provides the justification for crawler lanes on the up-hill gradients.

The design can therefore be seen to be logical in the context of severe economic constraints. Those responsible for preparing the design were faced with a dilemma: the road as built, with certain well understood shortcomings, or no road at all. Given the environmental benefits in the town High Street that resulted, most people would probably agree that they made the right choice.

The by-pass has not, however, been without its problems. Shortly after it opened, it became apparent that it was to have a bad accident record – not entirely surprising on a road known to be sub-standard for the traffic it was required to carry.

The accidents fell into two categories. Vehicles overtaking on the section of road that had no crawler lane provided hit vehicles coming in the opposite direction head-on, resulting in more than one fatality in the early months after opening. And vehicles approaching the terminal junctions down the steep gradient either failed to negotiate the tight curve leading to the minor road or were involved in a shunt accident because drivers failed to appreciate the slow speed of the vehicle in front.

The first of these categories says something about driver psychology and the desirability of crawler lanes on single carriageway roads in isolation from other factors of the design. In this particular situation, a long stretch of two lane road on a reasonable alignment but with little opportunity for overtaking is followed by a crawler lane for about 0.75 km and then a return to two lane road. With a high proportion of heavy goods vehicles in the traffic stream, it is relatively common for one HGV to overtake another through the whole of the length of the crawler lane. The impatient driver in the fast car is eager to overtake both HGVs so he remains in the outside lane beyond the end of the crawler lane, thus creating a conflict situation with vehicles coming in the opposite direction and, potentially, a very serious accident. There is not a great deal the designer can do to ease this situation except, perhaps, altering road markings and erecting warning signs. It can properly be described as a mistake initiated by a misguided search for economic rectitude. We should always try to learn from our mistakes.

The second category of accident tended to be less serious in terms of personal injury, but none the less highly undesirable. The original layout provided for short deceleration lanes at both terminal junctions. Traffic coming off the by-pass therefore had to slow down on the main carriageway on a downhill gradient and then negotiate a sharp lefthand

Box 9.1 continued:

---

curve to leave the main road. It was a difficult and possibly unexpected manoeuvre, and encouraged driver error. Some improvement could be implemented at low cost by simply increasing the length of the deceleration lane and this has been put into effect.

It is perhaps reassuring to note that the economic policies which led to this particular design have now been largely replaced. Perhaps lessons have been learned. However, the engineer should never see his job as simply implementing the will of others. Clients, even government departments, can be wrong, and when they are, engineers should not hesitate to tell them.

Selection of design speed

The design speed of a highway has been variously described by different writers. It does not have a precise meaning, and its purpose is to give guidance to designers in working out appropriate sight distances, transition lengths and superelevation. A number of factors influence the behaviour of drivers, and it is the function of the designer to combine these influences in such a way as to ensure that the maximum economic benefit consistent with safety is derived from the road.

A situation in which 85% of drivers travel at or below the design speed is generally regarded as being satisfactory. This will normally lead to a situation in which 99% of drivers travel at or below one standard speed band above the design speed. The standard speed band boundaries are 145 km/h, 120 km/h, 100 km/h, 85 km/h, 70 km/h, 60 km/h and 50 km/h. Below this level, design speed is not used as an input to alignment design except in the somewhat limited application on estate roads. Most rural roads have design speeds of 85, 100 or 120 km/hr.

On urban roads, the design speed is taken as being the speed limit which applies (an urban road is defined as being a road on which there is a speed limit of 50 mph or less).

On rural roads, the design speed is determined by the procedure outlined in TD9/81. This depends on three constraints, the statutory constraint, the layout constraint and the alignment constraint. A full explanation is shown in the advice note which accompanies TD9/81.

Statutory constraint refers to the general speed limit of 70 mph (60 mph on single carriageways) and these are approximately equal to

120 km/h and 100 km/h respectively. These are the maximum design speeds used.

In order to determine layout constraint and alignment constraint, it is necessary to sketch a trial alignment on a large scale (1/2500) Ordnance Survey sheet. The layout constraint, L_c, can then be determined from Table 9.2, where figures ranging from 0 (D3M) to 33 (6 m S2, High Access, Narrow Verge) can simply be read off from the table. There are circumstances in which the exact conditions identified in the table do not apply, and a sensible interpretation of the standard is required, possibly involving interpolation between adjacent figures.

Alignment constraint, A_c, is defined in TD9/81 as being

$$\text{Dual carriageways}: A_c = 6.6 + B/10 \qquad [9.1]$$

$$\text{Single carriageways}: A_c = 12 - VISI/60 + 2B/45 \qquad [9.2]$$

where B = Bendiness in °/km (See Figure 9.6); VISI = Harmonic Mean Visibility.

VISI has various values assigned to it in the standard for different situations. It can be estimated from the empirical relationship

$$\text{Log}_{10}VISI = 2.46 + VW/25 - B/400 \qquad [9.3]$$

where VW = Average verge width averaged for both sides of road; B = Bendiness in °/km.

Having used these relationships to determine A_c, you should then use Figure 9.4 to select the design speed. This figure is drawn from data collected from observations of driver behaviour on roads having certain values of L_c and A_c.

Table 9.2 Determination of layout constraint (Source TD9/81)

Road Type	S2		SW2		D2AP		D3AP		D2M	D3M
Carriageway Width (Ex. Metre strips)	6 m	7.3 m	10 m		Dual 7.3 m		Dual 11 m		Dual 7.3 m & Hard Shoulder	Dual 11 m & Hard Shoulder
Degree of Access and Junctions	H	M	M	L	M	L	M	L	L	L
Standard Verge Width	29	26	23	21	19	17	10	9	6	4	0
1.5 m Verge	31	28	25	23						
0.5 m Verge	33	30								

138

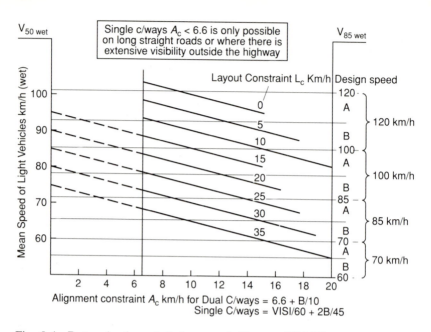

Fig. 9.4 Determination of design speed (Source: TD9/81)

It has been found that speed distributions remain reasonably constant across roads with similar geometric properties. For instance, studies of speed distributions on various roads have shown the results illustrated in Figure 9.5. This figure is reproduced from TA43/84 which establishes the relationship:

$$99\text{th }\%ile = 85\text{th }\%ile = \sqrt[4]{2} = 1.19 \qquad [9.5]$$

$$85\text{th }\%ile = 50\text{th }\%ile$$

This relationship suggests the adoption of a range of design speeds which are related to each other by a factor of 1.19. If we adopt the 85%ile speed as the design speed, and round the figures to give whole numbers, we can establish a structure for design speeds as shown in Table 9.3.

This allows us to make some reasonable judgements concerning the geometric properties of the road. If, by design speed, we mean the 85%ile speed, we can determine, from Table 9.3, what the 99%ile speed will be, and can select values for curvature, superelevation and sight distance which take account of this particular feature of driver behaviour. In practice, there are a number of points where 'desirable' and 'absolute' standards are defined. These imply that the 99%ile driver travelling round a 'desirable' curve will experience the same level of risk as the 85%ile driver travelling round a 'absolute' curve. The adoption of these design speeds, and the standards which follow

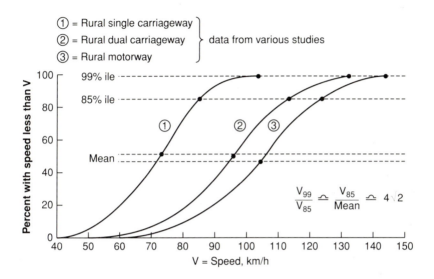

Fig. 9.5   Speed distribution curves on different categories of road (Source: TA43/84)

Table 9.3   Structured framework of design speeds (Source TA43/84)

50th %ile speed	85th %ile speed	99th %ile speed
100	120	145
85	100	120
70	85	100
60	70	85

from them, mean that we are aiming to design our road on the assumption that 99% of drivers are not travelling too fast. But in difficult locations, we can reduce that figure to 85%, or even 50% of drivers, although we may take other measures, such as signing, to influence driver behaviour.

It therefore follows that the designer can influence driver behaviour, i.e. speed, by selecting geometric properties appropriate to the circumstances. It may be that, by following through the procedure outlined, the designer finds that his alignment produces an 85%ile speed that is not appropriate to the circumstances, e.g. 120 km/h speed on a single carriageway or a theoretical design speed greater than 120 km/h or less than 85 km/h on an unrestricted road. There are a number of options that may be available in these circumstances: to change the alignment, which will have cost implications, or to introduce restrictions in the form of speed limits or advisory warning

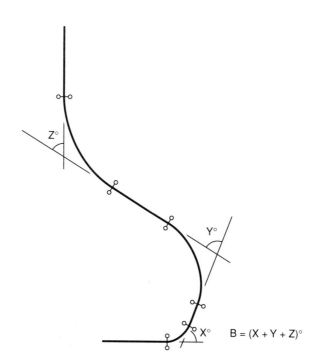

Fig. 9.6 Determination of bendiness

signs. These situations often arise on roads designed prior to 1981 which have since developed high accident rates or become congested, and the highway engineer's task is to find a means of improving the situation without necessarily incurring high costs of reconstruction.

TD9/81 therefore has applications other than those directly concerned with the design of new rural roads across green field sites, and it is important that the highway engineer should fully appreciate the relationship between the design of the highway and the behaviour of drivers who will subsequently use it.

Box 9.2:

This is an example of the way TD9/81 can be used to assess the performance of an existing length of highway.

A section of single carriageway rural road 3.5 km long constructed in 1968 between two roundabouts at A and B is being assessed to see whether marginal improvements are required to bring it up to modern standards. A sketch of the layout of the road is shown at Figure 9.7, and a cross section through the road at Figure 9.8. There are a total of 21 access points and minor junctions along the length of road under consideration.

Box 9.2 continued:

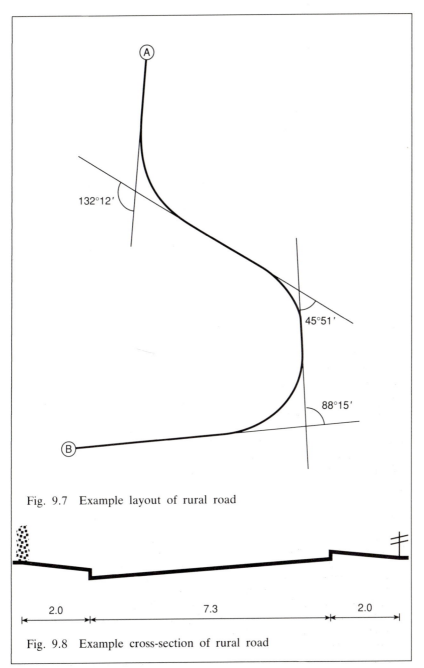

Fig. 9.7   Example layout of rural road

Fig. 9.8   Example cross-section of rural road

Box 9.2 continued:

The results of a spot speed survey are shown in the following table:

Speed Band (km/h)	Number of observations
<65	32
65–69	19
70–74	29
75–79	206
80–84	225
85–89	388
90–94	191
95–99	25
>99	13

Using the method outlined in Department of Transport Technical Standard TD9/81, and by calculating the 85th %ile observed speed, determine whether this section of road is appropriate for modern conditions and comment upon the improvements, if any, that should be made.

Analysis:

There are 21 access points in 3.5 km = 6/km
This is classified M
We have a S2 7.3 m carriageway with verge width between standard and 1.5
so, from Table 9.2 and interpolating $L_c = 24$
Bendiness (from Figure 9.7)
= [88°15′ + 45°51′ + (180°–132°12′)]/3.5
= 52°50′ = 52.833°
VW = 2.0
so substituting in equation 9.3
$\log_{10}$VISI = 2.46 + 2/25 – 52.833/400
= 2.41
VISI = 257.04
so substituting in equation 9.2
A_c = 12 – 257.04/60 + (2 × 52.833)/45
= 10.07

Thus, reading from Figure 9.1, nominal 'design' speed = 102 km/h
Observed 85th %ile speed is in the range 90–94 km/h

Conclusion:
We can thus see that drivers are actually travelling at a rather lower speed than the road design would allow for. This might mean that we could get relatively good value by carrying out minor improvements to sight lines, curvature etc, thus increasing the speed of drivers, reducing journey times and gaining economic benefits without risking an increase in the accident rate. It is probably more likely that we would leave well alone unless there were problems of congestion on adjacent parts of the network.

Box 9.2 continued:

> Had we got the opposite result, i.e. that drivers were travelling faster than the conditions allowed, we would probably have found that this was combined with a high accident rate and improvements would be more urgent.

References and further reading – Chapter 9

1. Department of Transport/Welsh Office. *Design Standard TD9/81. Highway Link Design.* Department of Transport. London. 1981. Also Amendment No 1. 1985.
2. Department of Transport/Welsh Office. *Technical Advice Note TA43/84. Highway Link Design.* Department of Transport. London. 1984.
3. Watson J P. *Highway Construction and Maintenance.* Longman Scientific and Technical. Harlow. 1989.
4. Department of Transport/Welsh Office. *Departmental Standard TD20/85. Traffic flows and carriageway width assessments.* Department of Transport. London. 1985.
5. Department of Transport. *Standard Highway Details.* HMSO. London. 1987.

Chapter 10

Highway Functions

The purpose of a highway

One of the first questions to which the designer of anything must find an answer is: 'what is its function?'

This is particularly important for the engineer because, by and large, engineers are concerned with products that are measured according to their usefulness. Most engineers would consider themselves to have failed if the outcome of their endeavours did not work.

Other professionals are not able to call upon such clearly definable objectives. Architects, for instance, are not normally willing to agree with the premise that their function is simply to make a building look nice while others, engineers, make sure it does not fall down. But they would agree that appearance, aesthetics, treatment of spaces and relationship with other components in the overall location are important and, what is more, their background and professional experience makes them particularly able to offer advice and pass judgement on such matters.

As engineers, we are usually concerned with more tangible concepts. We create a structure in order to perform some purpose, and if we cannot design it to fulfil this objective, then we do not create it.

A road, in this sense, is a structure, made of concrete, steel, macadam and other materials. If we are to design it, we must first identify its purpose, and this, in turn, means that we must identify its function or functions.

The addition of the 's' is important, for one of the problems faced by the highway engineer is that his creation will normally be required to perform a number of different, sometimes conflicting, functions. Other transport infrastructure does not often have the same problem. A railway line has no purpose other than to provide a track on which trains can run, and those trains are defined quite closely in terms of size, gauge, weight, speed etc. An airport runway is defined solely for the benefit of the aeroplanes that will land and take off from it and if, for instance, the runway is too short, particular types of aeroplane will never use it. A canal is designed in order to facilitate the passage of boats of a particular length, width and draught (a fact which fishermen might choose to dispute). A dock or a roll-on, roll-off (Ro–Ro) terminal is likewise limited to use by a particular size and type of ship, with the means of loading and unloading being defined too.

None of this applies to a road. Even a motorway, where the type of vehicle and driver are limited by law, must still cope with everything from a two-wheeled motor bike to a five-axled articulated truck, travelling at speeds between zero and seventy miles per hour, and more. Drivers, who have total control over their vehicle, may be fit, alert and highly skilled. Or they might be tired, depressed and suffering the after effects of too much lunch. The highway engineer must cope with a wide range of possibilities.

It is not possible to make out a list of highway functions that will apply to every situation. The motorway, considered above, is not very typical. The vast majority of roads are all purpose and are not subject to the legal constraints that are applied to motorways. We can, however, consider a range of possible functions from which the designer can select those which apply in a particular case. The engineer can then make some judgement regarding the relative importance of these various functions, and he has made a start on designing his road. The following list might describe likely functions of the highway:

To cater for through vehicular traffic

Much of the time we are using the highway as a means of getting from A to B and we will want to do this in safety, with reasonable speed and reasonable comfort. A great deal depends on exactly what we mean by the word 'reasonable'. In this country, Parliament has decided that reasonable speed entails a maximum of 70 miles per hour, although this is a matter of continuing debate. Lower speed limits are imposed because of conflicts which might occur with the needs of other users of the highway, or because it is unreasonable to suppose that the authorities will invest the resources necessary to make all roads suitable for 70 miles per hour running. From the point of view of the authorities managing the highway, there is a need to ensure the maximum flow of vehicles through the minimum road space. The provision of additional road space is expensive and, increasingly, politically impossible to bring about.

To provide access to frontage properties

It is very rare that vehicles are moving along the highway simply for the sake of moving along the highway. Most trips have an origin, a destination, and a journey purpose. The function of the highway is to provide a means of getting from the origin to the destination. These trip ends can be defined as being the points at which the highway part of the journey starts and finishes, i.e. all trip ends will be highway served.

Frequently, one or both ends of the trip will simply be a point on the highway. A delivery van travelling from a central bakery to a baker's shop will simply need to pull up on the highway outside the shop whilst the driver unloads. For many years now, it has been a condition implicit in planning consent that all new development should have adequate facilities off-street so that vehicles can pull up without obstructing passing traffic. But that is no help in areas where the existing buildings pre-date planning regulations. Even where service roads and off-street parking are available, frequent access points involving tight turning radii will have a major effect on the flow of traffic along a main road.

To provide a passageway for pedestrians

We are so concerned with the very obvious problems of congestion caused by too many vehicles in too little space that we tend to forget that by far the commonest mode of transport is our own two feet, at least in terms of numbers of trips, if not distance travelled. Walking is available to almost every member of the population, including the very old and the very young. The majority of serious injury and fatal road accidents occur to pedestrians, for obvious reasons. Part of the reason for the relative fall of number of accidents over the past twenty years must be the popularity of pedestrianisation schemes and the development of shopping malls, where pedestrian movement and vehicular movement are largely separated. But most highways still cater for both vehicles and pedestrians and even motorways, where pedestrian movement is generally prohibited, must still allow for maintenance workers, police and breakdowns.

In recent years, there has been something of a resurgence of interest in cycling as a means of transport and it has always been an important mode to those too young to drive. In the context we are considering here, cycling should perhaps be considered as an extension of walking although it is accepted that the speeds are higher and cyclists are generally banned from using the footway on urban streets.

To create a route for public transport

Whilst public transport is clearly vehicular, it should perhaps be considered on its own in respect of the function of the highway. An ordinary bus, operating on an all purpose road without any special facilities, has particular requirements in that it has to pick up and set down passengers. The location of bus stops therefore becomes an important feature of the design of the highway.

With increasing emphasis on maximising the use of public transport, the highway's function in this respect is likely to become more important. Bus lanes already exist on the radial routes leading into most major cities. Light rail rapid transit systems (LRT) operational or under construction in Manchester and Sheffield incorporate some sections of on-street running in order to avoid the massive expense and disruption involved in getting segregated LRT (e.g. London Docklands) into city centres. A further variation is the proposed guided busway in the York Road corridor in Leeds where buses would use the normal highway on the low traffic sections of their routes but the guideway, incorporated into the highway, in the congested sections.

To act as a car park

Car parking is now prohibited on most main roads in city centres, and loading is frequently limited by time of day. However, in residential areas, small towns, suburban shopping centres, old industrial areas and tourist centres the only place to park a car is on the street. Whilst it is an offence in British law to obstruct the highway, this would not generally be held to apply to parking on the street outside your own home unless specific regulations were in force. Many small retail businesses still perceive the availability of free parking close at hand as being essential to their trade, and if off-street parking is not available, this means that unreasonable restrictions to parking on-street would be disastrous. One function of the road is therefore perceived as being a car park.

To be a corridor for services

Despite the antipathy that has existed for many years between statutory undertakers and highway authorities, there is no real alternative to the use of the highway corridor for laying cables and pipes. It is to be hoped that the demise of the Public Utilities Street Works Act 1950, with its absurd and out-dated division of responsibility between authorities, and its replacement with the New Roads and Street Works Act 1991, will lead to closer understanding and co-operation.

To make a contribution to the townscape

Historically, towns came into being at locations where human activity took place. The location of the town was frequently determined by transport considerations. London was, to the Romans, the lowest crossing point on the Thames. First Bristol, then Liverpool became great cities because they were ports facing the right way. Many towns owe their origins to the fact that they were at a crossroad where farmers and traders could meet. The buildings, which we would now consider to be the primary feature of the town, were usually a secondary consideration.

It is perhaps because of the failure to appreciate the importance of transport as a land use in its own right that we are left with so many bleak and featureless town centres as a result of 1960s reconstruction. A town is, or should be, a coherent whole that is a centre for human activity. It consists of both buildings and transport systems, and there is not a great deal of point of having one without the other.

Arguably, we are learning from some of the mistakes of the past. There are a number of examples of townscapes that work well in both functional and aesthetic terms.

Capacity

In considering the function of a road to carry vehicular traffic, it will clearly be helpful if we can add some numbers to our definition. One way to do this is to look at capacity. Unfortunately, traffic is not an easy commodity to measure. It moves and its volume is not fixed, but varies with time. However, we can make a start by looking at the amount of space that is taken up by 1 unit of traffic.

Consider a vehicle 4.0 m long travelling along a two-lane dual carriageway. Effectively, it will be occupying one lane width, so we can consider the amount of road taken up as a length of road, understanding that it is actually an area. This has the advantage of removing one variable – the width of the vehicle – from consideration. For our present purpose this doesn't matter very much, for we are considering traffic moving freely along an open road. It will be more relevant when we come to consider the capacity of junctions.

Although a vehicle only occupies, instantaneously, a length of road equal to its own length, its presence means that a greater length of road is unavailable to any other vehicle. In order that the road can be used safely, there must be a gap, called the headway, between one vehicle and the next. This is illustrated in Figure 10.1. The length of road taken up by a vehicle is therefore equal to:

$$\text{Length (L)} + \text{Headway (H)}$$

$$\text{or } S = L + H \qquad [10.1]$$

where S is the road space utilised by the vehicle in metres.

If all the vehicles are travelling at speed V metres/second the time taken, t, for vehicle 2 to travel distance S and hence occupy the position of vehicle 1 is

$$t = S/V \text{ secs} = (L + H)/V \text{ secs} \qquad [10.2]$$

or, 1 vehicle passes an observer, O, every (L + H)/V secs, which could be said to be the physical capacity of a single lane. Putting some figures into this, we could look at vehicles 4.0 m long travelling at 100 km/h (27.8 m/sec) and headway of 50 m, we get t = 1.94 secs, i.e. a vehicle passes the observer every 1.94 secs. This in turn means that 1856 vehicles would pass the observer every hour, or the capacity of the lane is 1856 veh/hr.

Of course, not all vehicles are 4.0 m long, they do not all travel at the same speed and they do not always keep the right distance back from the vehicle in front.

The Highway Code[1] gives some guidance on appropriate headways when driving conditions are good. Values shown in Table 10.1 are taken from the Code.

From this table it is clear that stopping distances which the Code

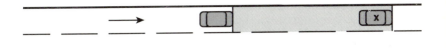

Car at **x** "occupies" shaded area

Fig. 10.1   The amount of roadspace 'occupied' by one vehicle

Table 10.1   Recommended Stopping Distance (Source: The Highway Code 1991)

Speed mph	Speed km/h	Distance m
20	32	12
30	48	23
40	64	36
50	80	53
60	96	73
70	112	96

recommends are appropriate headways, increase disproportionately with speed. The reason for this is that there is an element of 'thinking time', sometimes taken as two seconds, between the appearance of an obstacle and the brakes being applied.

It should also be clear from Table 10.1 that at relatively high speeds (80 km/h and above) the length of the vehicle is comparatively small when compared with the headway. Thus, on the dual two-lane all purpose road which we have been considering, the value, S, for a car travelling at 80 km/h is 53 + 4 = 57 m, and the equivalent figure for a 9.5 m long lorry is 53 + 9.5 = 62.5 m.

At 100 km/h (interpolating between values given in the table) headway becomes 79 m and S becomes 83 and 88.5 m respectively.

We can thus conclude that the capacity of a high speed road varies considerably with the speed at which vehicles are travelling, but does not vary much with the length of the vehicle. Given modern design of vehicles, where even the largest vehicles can travel at 100 km/h plus, it follows that the mix of vehicles in the traffic stream does not have a major effect on the capacity of the road. This works through into the design standards, where traffic flow is measured in vehicles per hour in order to determine the width, i.e. the number of lanes, that is required.

On low speed, i.e. urban, roads the same is not generally true. Comparing the value of S for a car and a lorry at 32 km/h we can establish values of 16 and 21.5 m respectively. The length of the vehicle accounts for between 25% and 44% of the total value of S. If vehicles slow to 5 km/h it might be reasonable to say that the distance between them is no more than 1.5 m. Our values of S have dropped to 5.5 m and 11 m and the length of the vehicle accounts for between 73% and 86% of the total space taken up. On low speed roads, therefore, we do have to take account of the mix of vehicles in the traffic stream. This is done by means of a weighting factor (the passenger car unit – pcu) and traffic flows are measured in terms of pcus/hour. The value of pcus that is applied to any particular vehicle varies with the circumstances. We will return to the subject of pcus in more detail when we consider Flow/Capacity ratios in the context of junction design in Chapter 12.

From the relationships shown above, it can be seen that, by reducing the headway without reducing the speed, the number of vehicles passing the observer, i.e. the capacity, increases. It is also possible to show that, by reducing the speed and maintaining the headway at recommended levels, capacity is increased to a certain optimum, and then it decreases. This is shown graphically in Figure 10.2. It therefore follows that to get the maximum capacity out of any particular section of road, all traffic should travel at the optimum speed, and this optimum speed is not the maximum legal speed. One argument against increasing speed limits on motorways above 70 mph is that, by so doing, you would actually decrease the capacity.

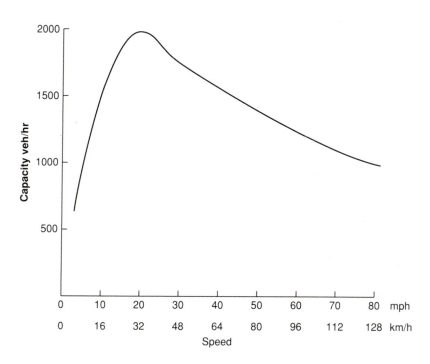

Fig. 10.2   Reduction of capacity with increased speed

Of course, the designer is not in a position to determine exactly how drivers will behave. It is, however, worth noting what happens when a motorway becomes congested. It is likely that the motorway will have a nominal design speed of 120 km/h and in the United Kingdom, the maximum legal speed is 70 mph (about 112 km/h).

References and further reading — Chapter 10

1. Department of Transport and Central Office of Information. *The Highway Code.* HMSO. 1991.

Chapter 11

Application of Design Standards

Highway link geometry

The starting point for determining the geometry of a highway is the design speed. The definition of design speed, and the method of determining its value in accordance with the Department of Transport Standard, TD9/81, is shown in Chapter 9.

Having selected a design speed, various other parameters associated with highway geometry can be read from Table 3 of TD9/81 (as amended), which is reproduced as Table 11.1.

As is discussed in Chapter 10, design speed bands of 50, 60, 70, 85, 100 and 120 km/h are used and are related to each other, approximately, by the $\sqrt[4]{2}$ relationship. In general, desirable and absolute minimum values are shown. Desirable minimum values are determined by comfort levels when a vehicle is travelling at the design speed, whereas absolute minimum values are the same as desirable minimum values for one design speed step above the design speed. There is also an additional limiting value for radius of horizontal curvature, which is the same as the desirable minimum value for two design speed steps above the design speed.

Faced with the selection of an alignment, the designer will initially try to ensure that desirable minimum standards are adopted throughout. However, a designer is rarely in a position to select an alignment

153

Table 11.1 Design parameters for horizontal curves (Source: TD9/81 amendment No 1)

Design speed km/h	120	100	85	70	60	50	V^2/R
A. Stopping Sight Distance m							
A1 Desirable Minimum	295	215	160	120	90	70	
A2 Absolute Minimum	215	160	120	90	70	50	
B. Horizontal Curvature m							
B1 Minimum R * without elimination of Adverse Camber and Transitions	2880	2040	1440	1020	720	510	5
B2 Minimum R * with Superelevation of 2.5%	2040	1440	1020	720	510	360	7.07
B3 Minimum R * with Superelevation of 3.5%	1440	1020	720	510	360	255	10
B4 Desirable Minimum R Superelevation of 5%	1020	720	510	360	255	180	14.14
B5 Absolute Minimum R Superelevation of 7%	720	510	360	255	180	127	20
B6 Limiting Radius Superelevation of 7% at sites of special difficulty (Category B Design Speeds only)	510	360	255	180	127	90	28.28
C. Vertical Curvature							
C1 FOSD Overtaking Crest K Value	*	400	285	200	142	100	
C2 Desirable Minimum * Crest K Value	182	100	55	30	17	10	
C3 Absolute Minimum Crest K Value	100	55	30	17	10	6.5	
C4 Absolute Minimum Sag K Value	37	26	20	20	13	9	
D. Overtaking sight distance							
D1 Full Overtaking Sight Distance FOSD m	*	580	490	410	345	290	

* Not recommended for use in the design of single carriageways (see Part C Paras 2.7–2.8)

regardless of other constraints. He will have to thread his way between obstacles, will have to ensure that structures are located at suitable positions, will have to optimise earthworks and will have to take account of environmental issues. Having established one or more feasible alignments, he will have to carry out an economic appraisal to demonstrate that his route provides value for money, and he may also have to contend with the practicalities of acquiring the necessary land and steer his way round political objections. So in reality, there may be circumstances in which it is not possible to meet all the desirable minimum standards all the time. Hence, the absolute minima and, in one case (line B6), the limiting value.

Sight distances

Lines A1, A2 and D1 of Table 11.1 show desirable and absolute value of stopping sight distance and the full overtaking sight distance (FOSD) for each of the standard values of design speed. These show the minimum distance that a driver should be able to see ahead according to whether the section of road is classified as an overtaking stretch or not. Overtaking sight distances only apply to single carriageways. There is no FOSD stated for a 120 km/h road as this design speed is not appropriate for a single carriageway.

The principle of stopping sight distance (SSD) is that a driver should always be able to stop in the distance that he can see ahead of him.

There are three components of SSD:

1. The distance travelled in the time taken to perceive the hazard – the perception distance.
2. The distance travelled during the time taken by the driver to apply the brakes and for the brakes to come into effect – the reaction distance.
3. The distance travelled whilst the brakes are applied to bring the vehicle to a stop – the braking distance.

The standard requires that an envelope of clear visibility should be provided as shown in Figure 11.1, such that an object between 0.26 m and 2.0 m above the road surface should be visible to a driver whose eye height is between 1.05 m and 2.0 m.

Since the hazard relevant to FOSD will normally be a vehicle coming in the opposite direction, it is not necessary to consider an object height close to the road surface. The envelope of clear visibility is therefore that shown in Figure 11.2. Note that these provisions allow for obstructions above the carriageway such as overhanging trees or bridge soffits.

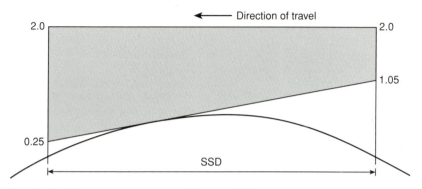

Fig. 11.1 Envelope of clear visibility for SSD

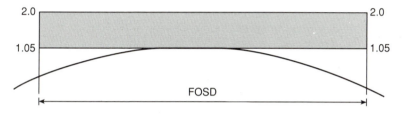

Fig. 11.2 Envelope of clear visibility for FOSD

Clearly, perception and reaction time can vary widely from driver to driver, and even for the same driver depending on his state of health, degree of tiredness or whether he has been drinking. In determining SSDs, a value of two seconds has been adopted, which is rather longer than has been established in various tests.

The rate of retardation of the vehicle will depend on the friction developed between the tyre and the road surface and the efficiency of the braking system. A value of 0.25 g can be developed without causing discomfort to the passengers or the risk of skidding taking place, so the distance travelled in

$$2 \text{ seconds} + V/0.25g \qquad [11.1]$$

is taken as the desirable minimum SSD. (V = Design Speed).

If we adopt the suggestion that the absolute minimum value should be the same as the desirable minimum for one design speed lower, then we get absolute minimum SSD as the distance travelled in

$$2 \text{ seconds} + V'/0.25g \qquad [11.2]$$

where V' is one design speed step below V. This is approximately the same as

$$2 \text{ seconds} + V/0.375g \qquad [11.3]$$

A rate of retardation of 0.375 g is perfectly acceptable, although it may lead to some passenger discomfort.

FOSD in Table 11.1 refers to full overtaking sight distance. Ideally, on a single carriageway road, FOSD should be provided on sections specifically designed to permit overtaking.

It is not possible to be precise about the exact nature of an overtaking manoeuvre, although studies have shown that the time taken varies from about 4 seconds to about 15 seconds and is largely independent of vehicle speed.[1] Eighty-five per cent of overtaking manoeuvres have been observed to take less than 10 secs and 99% less than 14 seconds.

FOSD is considered to be made up of three components, D_1, D_2 and D_3, which are shown in Figure 11.3. D_1 represents the distance travelled by the vehicle doing the overtaking during the overtaking manoeuvre; D_2 represents the distance travelled by a vehicle coming in the opposite direction; and D_3 represents a safety factor.

For establishing design values of FOSD it is assumed that the overtaking driver commences the manoeuvre at two design speed steps below the design speed and accelerates to the design speed. The approaching vehicle travels at the design speed. D_3 is taken to be $D_2/5$. These assumptions give

$$\text{FOSD} = 2.05 \times \text{T} \times \text{V}' \text{ (metres)} \qquad [11.4]$$

where T = Time to complete manoeuvre (secs); V' = Design speed (m/sec).

Thus, if we want to cater for the 85%ile driver on a road having a design speed 100 km/h (27.78 m/sec), we can say

$$\text{FOSD}^{100} = 2.05 \times 10 \times 27.78$$

$$= 569.49 \text{ m}$$

We can obtain a very similar result by going back to the original assumptions that we made about design speed. We have already said:

D_1 = distance that the overtaking vehicle travels during the overtaking manoeuvre, assuming that it commences overtak-

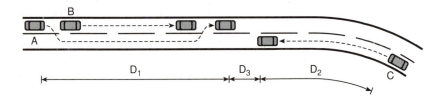

Fig. 11.3 Components of overtaking sight distance

ing at 2 design speed bands below the design speed and accelerates uniformly reaching the design speed as the overtaking manoeuvre is completed

D_2 = distance travelled by the opposing vehicle, assuming that it is travelling at the design speed

$D_3 = D_2/5$

Say the design speed = 100 km/h and average speed of overtaking vehicle is 85 km/h.

In 10 secs overtaking vehicle travels $(10 \times 85 \times 1000)/3600$ m
$$= 236 \text{ m}$$
$$= D_1$$
Oncoming vehicle travels $(10 \times 100 \times 1000)/3600$ m
$$= 278 \text{ m}$$
$$= D_2$$
Safety margin (D_3) $\qquad = 278/5$ m
$$= 56 \text{ m}$$
FOSD $= D_1 + D_2 + D_3 = 236 + 278 + 56 = 570$ m

The figure shown in Table 11.1 is marginally higher, 580 m.

It is important that the designer ensures that there is no confusion in the driver's mind as to whether it is safe to overtake or not. On sections of road designated for overtaking, sight distances should be at least the FOSD, although some limit will be necessary if the designer is to guard against long straight stretches of road which are boring to the driver and encourage high speeds. Where the section is not designated for overtaking, care should be taken to ensure that this fact is apparent to all drivers. If it is not possible to achieve FOSD, therefore, sight distance should be restricted to something not very much greater than the SSD.

Horizontal alignment

The first stage in fixing the geometric parameters to which a road is to be set out is to determine the shape of a reference line, usually the centre line, along the length of the road. There are two basic alignments that can be used: arcs and tangents and the cubic spline. The cubic spline is too complicated to work out manually, so we will concentrate here on arcs and tangents. We will return to the cubic spline in Chapter 14, where we are concerned with computer-aided design.

The basic form of a horizontal alignment is shown in Figure 11.4. A curve, which consists of a horizontal arc and two transition curves

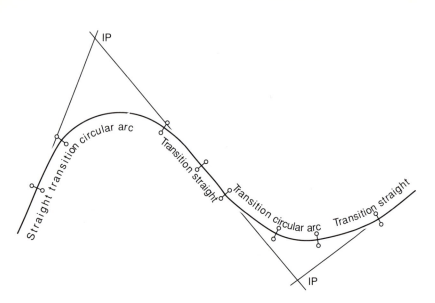

Fig. 11.4 Form of horizontal alignment using arcs and tangents

(spirals), joins two straights, which intersect at an angle I. In certain special circumstances, the transitions may be considered to have zero length, which explains the situation in Line B1 of Table 11.1 – minimum R without . . . transitions.

The procedure for fixing the curve begins with the definition of the two straights which it will join. The designer may have freedom of action in doing this, or he may be constrained by physical features through which the road will pass or by the end conditions where a new section of road joins an existing one. Having defined the straights, it is necessary to measure the angle between them – the intersection angle.

Reference to Table 11.1 will tell the designer the minimum radius that he can use for a given design speed. The minimum radii shown in the table follow from the requirement that superelevation on a public road should not be steeper than 7% in order to accommodate the needs, not only of traffic travelling at the design speed, but also very slow moving traffic such as bicycles.

A diagram of the forces acting on a vehicle of weight W as it travels round a bend of radius R is shown in Figure 11.5.

Resolving these forces at right angles to the surface of the road, we have

$$(Mv^2/R)\cos \alpha = Mg\sin \alpha + P \qquad [11.5]$$

and resolving parallel to the surface of the road

$$P = \mu [Mg\cos \alpha + (Mv^2/R)\sin \alpha] \qquad [11.6]$$

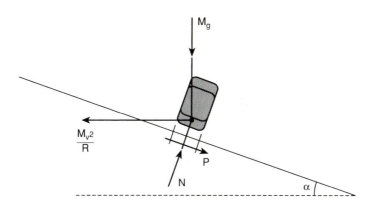

Fig. 11.5   Forces on a vehicle travelling round a horizontal curve

Thus

$$v^2/gR = \tan \alpha + \mu + (\mu\,v^2/gR)\tan \alpha \qquad [11.7]$$

If we ignore the last term of this expression as being very small and express the angle of superelevation α in terms of the slope e, then

$$v^2/gR = e + \mu \qquad [11.8]$$

If we now replace v m/sec with a speed V expressed in km/h, and put g = 9.81 m/sec,

$$V^2/127R = e + \mu \qquad [11.9]$$

In applying superelevation, account is taken of the mean speed of vehicles, which, in turn, is taken as 67% of the design speed. No account is taken of the value of μ. The above equation then becomes

$$(0.67V)^2/127R = e = V^2/283R \qquad [11.10]$$

or, expressing e as a percentage

$$e\ (\%) = 0.353V^2/R \qquad [11.11]$$

which is the recommended formula for determination of superelevation, from which the values of R in Table 11.1 are derived.

The designer's next task is to determine which of the infinite number of designs will best satisfy his requirements, given the physical constraints and the need to optimise the economics of the design. He may already have an outline, sketched on a 1/2500 or 1/10000 ordnance sheet, which was needed in order to determine the design speed. He now needs to work up this outline design in order to specify the three dimensional co-ordinates of the finished road.

His first task is to consider the horizontal alignment of his reference line, which he can do by applying various geometric formulae. With reference to Figure 11.4, we know the position of the intersection point, I, and the direction of the two straights IT_1 and IT_2. These can be expressed by a combination of co-ordinates and bearings. Thus, if we can establish the length of the tangent $IT = IT_1 = IT_2$, we will know the position of the points T_1 and T_2. The appropriate formula is

$$IT = L/2 + (R + S)\tan(\alpha/2) \qquad [11.12]$$

where L = Length of transition; R = Radius of circular arc; S = Shift; α = Intersection angle.

We need two further formulae in order to solve this equation:

$$S = L^2/24R \qquad [11.13]$$

$$L = V^3/46.7qR \qquad [11.14]$$

where V = Design speed (km/h); q = Rate of change of radial acceleration of the vehicle as it passes along the transition (m/sec³).

These formulae are somewhat awkward and it is important to remember that the convention in Britain is always to express Design Speed in km/h, even though it may share a formula with another variable measured in metres or m/sec.

The proofs of these formulae follow from the geometric properties of the transition. Further information can be found in reference 2 or in surveying textbooks.

The purpose of the transition is to allow the gradual introduction of centrifugal forces which are necessary to cause the vehicle to pass round a circular arc, as opposed to continuing in a straight line. The driver will take a certain finite amount of time to turn the steering wheel, and, from the point of view of both comfort and safety, this period of time should not be too short. As the driver turns the wheel, the vehicle will follow its own transition curve. At slow speeds and large radii, it may not be necessary to take account of this in determining the layout of the highway, but at high speeds or on sharp radii, failure to incorporate the detail of the vehicle's path into the geometry of the highway can lead to slow moving vehicles tending to drift towards the centre of the arc, i.e. into the path of on-coming vehicles. The minimum radius for which it is not necessary to use transitions for various design speeds is shown in line B1 of Table 11.1, but this excludes the vast majority of curves designed in Britain.

The appropriate value of q, rate of change of radial acceleration, is a matter of choice for the designer. It is inappropriate for it to rise above a value of 0.6 m/sec³, as this can lead to instability of the vehicle. It is normally felt that it should not fall below 0.3 m/sec³ as this will tend to lead to very long transition curves and likely problems with overcoming geometric constraints. Thus, a designer will normally start with a value of q = 0.3 m/sec³, but if this proves difficult to incorporate on the ground, will allow it to rise.

Having determined the points T_1 and T_2, it is necessary to establish a series of points along the transition and circular arc. Two formulae are available for this purpose, depending on the precise form of spiral to be used. This, in turn, depends upon the intended use of the information obtained. There is not much practical difference between the two curves.

If the information is to be used to set out a curve using a theodolite and steel tape, the Euler Spiral is the most convenient form of the transition. With the theodolite set over the tangent point T_1 (or T_2), a series of angles δ_1, δ_2 . . . δ_n are set off from the tangent, as shown in Figure 11.6(a) and a series of chord lengths L_1, L_2 . . . L_n are measured. The value of angle to be set off for each chord length is given by the formula

$$\delta_n = l_n^2/2RL \qquad [11.15]$$

If points on the curve are required in terms of co-ordinates, it is more convenient to use a cubic parabola, as shown in Figure 11.6(b). In this case, the transition is defined by the equation

$$x = y^3/6RL \qquad [11.16]$$

This gives $x-$ and $y-$ co-ordinates relative to an origin at T_1 (or T_2) and with axes along and perpendicular to the tangent. It is, however, relatively straightforward using modern computer technology to rotate axes and shift origins so as to express co-ordinates in any convenient form. Computer systems such as MOSS will do this automatically.

Both these equations involve certain approximations, but these are not significant in the vast majority of cases. In the event of having to design a transition which is particularly awkward, either because it is longer than would normally be expected or because there are physical constraints present which restrict the designer's options, reference should be made to the curve ranging section of a specialist surveying text book.

One practical method for the designer to adopt if a computer is not available, is to work from a drawing using railway curves, transition templates and tables published by the County Surveyors' Society.[3] These adopt standard values of q = 0.3, 0.45 and 0.6 m/sec[3] and give a wide range of alternatives for R and V.

A fully worked example of the determination of the horiziontal alignment of a section of highway is included at Appendix B.

Vertical alignment

Having fixed the proposed horizontal alignment of a suitable reference line, the designer's next task is to fix the levels of the same reference line.

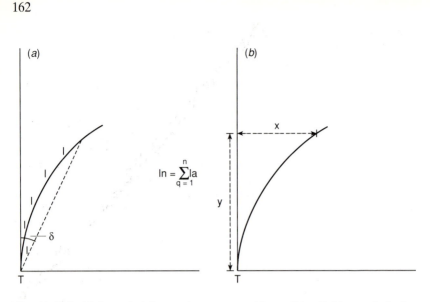

(a)

$$\ln = \sum_{q=1}^{n} la$$

(b)

Fig. 11.6(a)   Euler spiral for setting out transitions; (b)   Cubic parabola for setting out transitions

The basic form of a vertical alignment is made up from a series of straight gradients joined by vertical curves, which are normally parabolas, as illustrated in Figure 11.7. The geometry of the curve follows from its length, which in turn is a function of design speed.

Table 11.1 gives a series of values for 'K' dependent on design

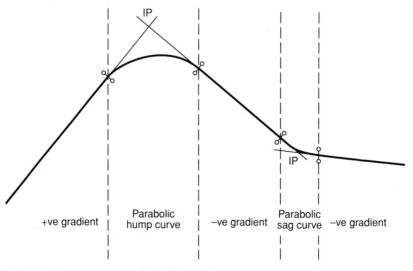

+ve gradient | Parabolic hump curve | −ve gradient | Parabolic sag curve | −ve gradient

Fig. 11.7   Form of vertical alignment

speed, whether the curve is a crest or a sag and whether overtaking or stopping sight distances apply.

The required length of the curve is given by

$$L = K(p - q) \qquad [11.17]$$

where K = value taken from Table 11.1; p,q are gradients expressed as percentages with an appropriate sign convention, e.g. a rising gradient from low chainage to high chainage is +ve, a falling gradient is −ve.

Table 11.1 gives us four alternative K values, according to whether we want Full Overtaking Sight Distance (FOSD), desirable minimum crest value, absolute minimum crest value or sag value.

The values of p and q vary according to the topography of the land across which the road is to be constructed. A minimum longitudinal gradient on a road of 0.5% (1 in 200) is normally reckoned to be the minimum permissible to ensure run-off of surface water. Where the road crosses land that is flatter than this, false drainage is incorporated into the longitudinal profile, such that the road rises at 0.5%, then falls at 0.5%, then rises again. (See Figure 11.8). An alternative approach is to ensure that there is continuous drainage to a French drain, or similar, running alongside the pavement and adjacent to it.

There are various formulae available to determine the levels along the curve which can be derived from the equation of a parabola. For the derivation of this formula, see reference 2, pp. 393–4 of O'Flaherty's *Highways* or any surveying text.

The formula used here is

$$y = [(q - p)/2L]x^2$$

where q, p and L are as before; x is the distance along the curve measured from the start of the curve; y is the vertical offset from the continuation of the gradient to the curve.

These parameters are illustrated in Figure 11.9.

An example of the calculation of levels along a vertical alignment is shown in Appendix C.

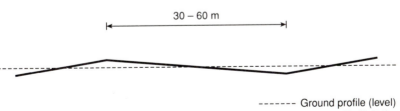

Fig. 11.8   Longitudinal section with false drainage

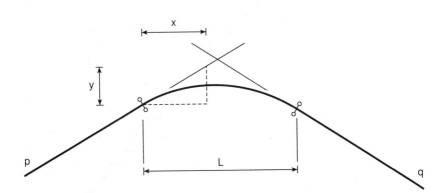

Fig. 11.9   Parameters used for levels on vertical curves

Crossfall, superelevation and the road in cross-section

The minimum fall across a road is 2.5% (1 in 40) to ensure that water falling on the road surface runs quickly to the channel and hence out of the wheelpaths of vehicles, where spray can be a serious inconvenience and danger. On a straight section of road this crossfall can be provided either as a straight fall from one edge of the pavement to the other (see Figure 11.10(a)), or as a 'balanced' road, with a fall from the centre to both channels (see Figure 11.10(b)).

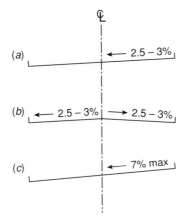

Fig. 11.10 (a)   Cross-section of road with crossfall; (b)   Cross-section of 'balanced' road; (c)   Cross-section of road with superelevation

On horizontal curves, however, superelevation must be applied to assist vehicles passing round the curve (see Figure 11.10(c)).

We have already seen, when considering the development of minimum radii of curvature shown in Table 11.1 that the amount of superelevation to be applied to a curve of radius R on a road with a design speed V is given by

$$e\ (\%)\ =\ 0.353V^2/R \qquad [11.17]$$

We now need to consider the cross-section of the road in order that we can determine the level at the edge of the carriageway, which is probably what we will need in order to determine the total amount of earthworks to be carried out, and to set out the construction of the pavements.

Various standard alternatives are available for the cross-section of a rural road in the UK, and these are shown in the Department of Transport Standard Details.[4]

For instance, Figure 11.11 shows some of the details given on Drawing A2 of the standard construction details relating to a single two-lane carriageway (S2). The standard lane width is 3.65 m, and in rural areas, edge treatment consists of a 1.0 m strip of the same construction of the remainder of the carriageway and delineated by a solid white line. The full width of the pavement structure is therefore 9.30 m. There is then an additional 2.50 m of verge, in which drainage and services may be located. The verge may be wider if visibility requirements dictate, or if additional services or signing are required. The level of the verge will be joined to the level of the adjacent ground

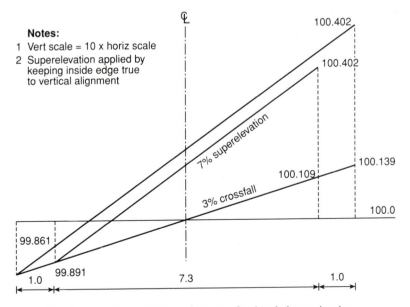

Fig. 11.11   Cross-section of S2 carriageway for level determination

by a batter, the steepness of which will vary according to the nature of the soil but may, typically, be 50% (1 in 2) to 25% (1 in 4). There will also be a narrow berm between the top or bottom of the batter and the fence erected on the highway boundary. In deep cuttings or high embankments additional berm(s) may be constructed part of the way up the batter, depending on the requirements for stability of the slope.

It is, therefore, not until the designer has gone right through the various stages of alignment design, as well as gathering information regarding soil reports, that he is able to fully define the area of land that the road is going to take up. At particular points, there may be physical constraints such as buildings or natural features, that impinge on the edge of the highway created by a particular alignment, and these problems may not become apparent until the cross-sections are drawn out. In these circumstances, the designer has the alternative of going back to the beginning and starting again on an amended alignment, or introducing some new feature, such as a retaining wall, to keep the whole of the highway within available space.

Even if, at the end of several trial runs, the designer finds that he has a feasible scheme, it does not necessarily follow that it is the best scheme. The criterion against which any design is measured is economic return. Any change in the design will inevitably affect the cost, but it may also affect the benefits as well. Reducing a gradient will affect the speed at which traffic travels on the road, particularly uphill. It will therefore affect time savings and hence, through the COBA analysis, benefits attributable to the the new road.

It is part of the skill of an experienced designer to be able to foresee problems before they arise. But even an experienced designer is likely to need several runs through the procedure before arriving at a design that satisfies all the constraints. Working manually, from tables and using a calculator and a drawing board, this is a very time consuming process.

It is here that the application of Computer Aided Design comes into its own. The most common system of CAD for major highways is MOSS,[5] which is a three dimensional surface modelling system. MOSS is more fully described in Chapter 14, but it is appropriate here to comment on the major advantages of CAD when compared with manual methods.

The latest version of MOSS is 9.0 in which most instructions to the computer can be given interactively by moving a cursor on a graphics screen. It is necessary to work through the process of building up a horizontal and vertical alignment in a similar way to that undertaken manually. Once the alignment is defined, however, it is stored in the computer and can be represented graphically on the screen. A situation is which the toe of the embankment impinges on an area of land which is not available to the highway, for instance, which might mean another three days work on the drawing board, is easily rectified. The right position for the embankment toe can be identified visually on the screen and the program will, very quickly, undertake all the necessary modifications to the entire alignment.

The main advantage of a computer, in highway design as with everything else, is that it can undertake a vast number of very simple calculations very quickly. Thus, in highway alignment design, which is basically a trial and error process, a computer system can arrive at an optimum solution far more quickly than a designer working manually. Because the process is speeded up so dramatically, more trials become feasible, and the final version should therefore be of a higher standard.

Box 11.1:

From the relationships developed in the section on sight distances, page 155, we can calculate minimum stopping sight distances. As an example, consider a road having a design speed of 100 km/h. We should find that the desirable minimum SSD is the same as the absolute minimum SSD for a road having a design speed of 120 km/h.

Solution:

Step 1: Express design speeds in term of m/sec

$$100 \text{ km/h} = (100 \times 1000)/3600 \text{ m/sec}$$

$$= 27.78 \text{ m/sec}$$

$$120 \text{ km/h} = 33.33 \text{ m/sec}$$

Step 2: Determine time taken to come to a halt

From 100 km/h @ 0.25 g

$$t = 27.78/(0.25 \times 9.81)$$

$$= 11.33 \text{ secs}$$

From 120 km/h @ 0.375 g

$$t = 33.33/(0.375 \times 9.81)$$

$$= 9.06 \text{ secs}$$

Step 3: Determine the perception/reaction distance

2 secs @ 100 km/h

$$d = 55.56 \text{ m}$$

2 secs @ 120 km/h

$$d = 66.66 \text{ m}$$

Step 4: Determine braking distance

From 100 km/h in 11.33 secs

$$d = 11.33 \times (27.78/2) = 157.37 \text{ m}$$

Box 11.1 continued:

From 120 km/h in 9.06 secs

$$d = 9.06 \times (33.33/2)$$

$$= 150.98 \text{ m}$$

Step 5: Add the distances from Step 3 and Step 4

Desirable SSD from 100 km/h

$$= 55.6 + 157.37$$

$$= 212.97 \text{ m}$$

Absolute SSD from 120 km/h

$$= 66.66 + 150.98$$

$$= 217.64 \text{ m}$$

References to Table 11.1 above shows the design value for both these parameters to be taken as 215 m.

References and further reading — Chapter 11

1. Transport and Road Research Laboratory. *A study of overtaking on rural trunk roads.* Unpublished. 1978. Quoted in Department of Transport. Advice Note TA43/84. DTp. London. 1984.
2. O'Flaherty C A. *Highways – Volume 1.* 3rd Edition. 385–389. Edward Arnold. 1986.
3. County Surveyors Society. *Table of properties of transition curves (metric).* CSS. London. 1974.
4. Department of Transport. *Standard Construction Details.* HMSO. London. 1987.
5. MOSS Systems Ltd. *MOSS V9.0.* Horsham. 1991.

Chapter 12

Junction Design

Types of highway junction

The function of a highway junction is to control merging and conflicting traffic streams in safety and with a minimum of delay. This it does by means of geometric layout, which controls vehicle paths through the junction, and regulation, which determines priority.

Highway junctions can be classified by layout type:

1. Major/minor junctions, sometimes referred to as priority intersections.
2. Signal controlled junctions.
3. Roundabouts, including mini-roundabouts.
4. Grade separated interchanges.

More than one of these layout types may be incorporated in a single junction. For instance, the layout shown in Figure 12.1 incorporates (b), (c) and (d).

Regulations are normally determined by Ministerial orders, which, in turn, are empowered by Acts of Parliament, usually, as far as road junctions are concerned, the Road Traffic Acts, most recently, the Road Traffic Act 1991.

The implementation of a Ministerial Order by the highway authority may require a specific decision by the elected members. In order to determine priority at junctions, however, it is a matter for the highway engineer responsible for the design of the junction to provide signs and road markings, and ensure that the layout of the junction

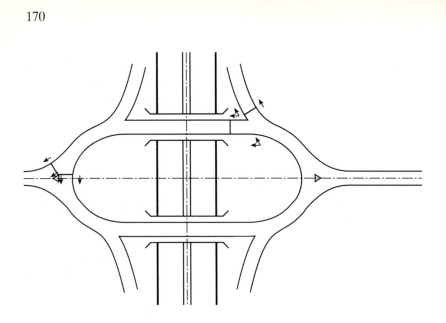

Fig. 12.1   Highway junction incorporating grade separation, roundabout and signal control

conforms with the standard pattern contained in the Traffic Signs Manual.[1,2,3,4] The Manual lays down the design of all road signs and markings in very great detail so that, for instance, the length and width of each section of white line in a 'Give Way' marking is prescribed.

The layout of a junction, be it major/minor, roundabout, signal controlled or grade separated is a matter for the design engineer to determine. It is his task to combine the various parameters so as to arrange a layout that is practical, meets the demands of the traffic, is safe and cost effective. Every road junction is different as it will be in a different location and matters such as the quantity of earthworks involved will have to be determined if a cost effective solution is to be arrived at.

Size of junctions

The starting point for any junction design is the quantity of traffic that is going to use it and the various turning, conflicting and merging movements that are involved. Thus, as with many aspects of highway design, we need an estimate of the traffic flows at some stage in the future, which we are going to use as the basis of our design. In the case of junctions, this is called the Design Reference Flow (DRF). Determination of DRF involves all the uncertainties which accompany any other process of traffic forecasting. British practice therefore allows for a range of DRFs to be generated from the traffic forecasting process, and alternative designs developed. Each of these is then evaluated in terms of their economic, environmental and operational impacts and a choice between them is made.

The detailed procedure for the determination and use of a range of DRFs is shown in Department of Transport Technical Memorandum TA23/81 entitled 'The Size of Junctions'.[5]

DRFs are defined as being peak hourly flows, but the question arises, which peak? If the junction is to be designed on the basis of traffic flows likely to occur 10 to 15 years into the future – a likely situation given the need for economic and environmental viability – it is unlikely that the absolute maximum flow that will occur in that time will be suitable for use as DRF. It may be that the Annual Average Hourly Flow (AAHT) is the appropriate figure to use, where:

$$AAHT = AADT/24 = (Annual\ Flow)/8780$$

(See Chapter 7 on Traffic Forecasting and Traffic Appraisal Manual.[6])

To some extent, the flow that is likely to prove economically and environmentally justified will depend upon the function of the road in question. Thus, on a recreational road, where very high peaks occur but only infrequently, the 200th highest flow in the life of the junction may be the one which gives economical viability. In urban situations, where peaks are much less marked, the 30th highest flow may be appropriate, whereas on 'typical' inter-urban roads, the 50th highest flow could apply. The implication is that, where the design flow is exceeded in practice, some degree of congestion will occur, but it is considered uneconomic, and probably environmentally undesirable, to design for a situation in which congestion never occurs.

There is a difference in the approach required according to whether the junction being designed is an improvement to an existing junction, or whether it is entirely new. If there is an existing junction, movement through which is not likely to change substantially due to new developments, base data for DRFs can be obtained from surveys. If, however, the junction is entirely new – the terminals on a by-pass, for instance – then the DRFs can be derived from the traffic forecasting model that is used in the design of the road.

A relatively common situation is a halfway stage between these two situations: where a junction is being provided on an existing road to provide access to a new development. In this situation, the design of the junction is a matter for agreement between the promoter of the development and the local highway authority. The developer clearly requires that the junction will function efficiently and, at least with certain types of development, will tend to attract potential customers and users. On the other hand, the developer will want to minimise costs as the junction will not directly influence income. It is likely that the highway authority will eventually 'adopt' the junction as part of the public network, and it will therefore be able to require that it is constructed to certain standards. The authority will be concerned to ensure that the junction does not contribute to unacceptable accident rates, does not cause delay on the rest of the public network and does not give it excessive liabilities for future maintenance expenditure. Local authorities have powers to impose their will through the procedures for adoption and, in cases where the highway authority is also the planning authority, through the need for them to give planning consent. In general, however, local authorities favour development in their areas because it constitutes inward investment and creates employment and other economic activity. So there is a balance to be struck. It is in everybody's interests to ensure that sensible agreements are reached between developers and local authorities, and both sides will have to call upon the traffic engineer's expertise. One likely area for compromise is the design of the junction giving access to the development.

Having determined the DRF for the junction, the next stage is to develop alternative junction layouts, either manually on a drawing board or using a computer-aided draughting system. Some CAD packages have typical junction layouts built into them which are suitable for use in relatively simple situations, such as a junction between two estate roads. In general, however, the junctions are the most expensive parts of a new road system, and particular attention to detail is required if the eventual design is to be cost effective.

Major/minor junctions

The simplest form of junction is the major/minor or priority intersection, and guidance on the layout, together with some 28 example drawings are shown in Department of Transport Technical Memorandum TA20/81.[7]

The basic principle of major/minor junction design is that it should follow the pattern of traffic movement, and that the heaviest flows should be given the easiest paths. Particular attention should be paid to situations in which there are likely to be significant amounts of heavy goods vehicles, particularly where they are involved in turning movements. Conflicts between traffic streams can be separated by either physical or ghost islands (diagonal hatching painted on the carriageway), although it is important that the driver's path through the junction should be clear well in advance. Large numbers of small islands tend to be confusing and therefore ineffective.

Visibility is crucial to the layout of a major/minor junction, as priority is always with traffic on the major road, and minor road traffic will therefore have to identify gaps in the major stream before being able to move. At worst, poor visibility will lead to an unacceptably high accident record, and at best it will seriously reduce the capacity of the junction. Visibility is often dependent on levels adopted, and the determination of any 'standard' design, applicable in any situation, is therefore impossible.

Traffic seeking to move from the major road to the minor will either have to slow down while moving along the major road in order to turn left, or will have to slow down and possibly stop in order to turn right. The presence of diverging lanes will allow this to happen safely and without causing delay to through traffic on the major road. The length of diverging lanes required will depend on the amount of turning traffic and, for right-turners, the amount of traffic in the opposing flow. Figure 12.2 shows a typical arrangement for a major/minor junction with diverging lane for right-turning traffic.

In certain circumstances, merging lanes can be provided to allow traffic from the minor road to accelerate before joining the major traffic stream, although these can be confusing. The driver has to successfully gauge the speed of the traffic stream he is about to join, and he only has the length of the merging lane in which to do it. Except in the situation of a full size merging lane as part of a grade separated junction on a motorway or other high speed road, it may be safest to avoid acceleration lanes altogether and force the driver to wait for a suitable gap in the oncoming traffic before proceeding.

There are three basic types of major/minor junction for single carriageways:

1. Simple
2. Ghost island
3. Single lane dualling.

The principles can also be applied at dual carriageway sites.

Junctions can be in the form of a T-junction, or a staggered junction (effectively, two T-junctions on opposite sides of the road designed in conjunction with each other). Crossroads should be avoided wherever possible as they invite drivers on the minor road to misunderstand the priorities, and, consequently, have a bad accident record. Since accidents at crossroads tend to be side on, and vehicles generally give less protection than in a head-on or shunt collision, injury rates are also bad.

Simple junctions are normally appropriate for accesses and minor roads in urban or rural areas, where the flow on the minor road is expected to be less than 300 veh/day 2-way AADT, or where there are serious physical constraints on the amount of land available and traffic on the major road is subject to a speed restriction of 40 mph or less. A simple junction consists of a T-junction without ghost or

174

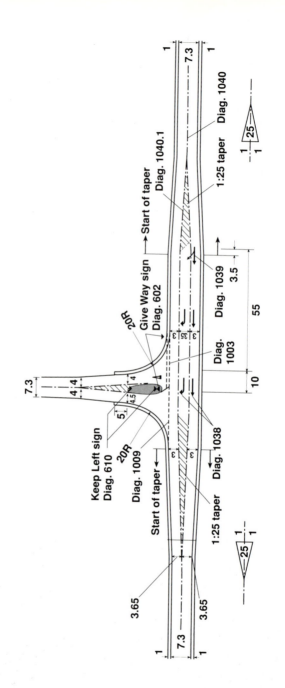

Fig. 12.2 Major/minor junction layout with diverging lane for right turning traffic (Source TA20/81)

physical islands. Wherever possible, the minor road should be at right angles to the major.

Having determined a suitable layout, next it is necessary to determine the capacity for each movement through the junction. These depend on the geometric properties of the junction and the quantity of traffic in the conflicting and merging flows.

Predictive equations for turning stream capacities are given in TA23/81, and are reproduced below. The terminology is illustrated in Figure 12.5.

$$qs_{b-a} = D\{627 + 14W_{cr} - Y[0.364q_{a-c} + 0.114q_{a-b} + \\ 0.229q_{c-a} + 0.520q_{c-b}]\} \qquad [12.1]$$

$$qs_{b-c} = E(745 - Y[0.364q_{a-c} + 0.144q_{a-b}]) \qquad [12.2]$$

$$qs_{c-b} = F(745 - 0.364Y[q_{a-c} + q_{a-b}]) \qquad [12.3]$$

where

$$Y = (1 - 0.0345W) \qquad [12.4]$$

and the parameters D, E and F refer to particular routes through the junction and are given by

$$D = [1 + 0.094(w_{b-a} - 3.65)][1 + 0.0009(Vr_{b-a} - 120)] \\ [1 + 0.0006(Vl_{b-a} - 150)] \qquad [12.5]$$

$$E = [1 + 0.094(w_{b-c} - 3.65)][1 + 0.0009(Vr_{b-c} - 120)][12.6]$$

$$F = [1 + 0.094(w_{b-c} - 3.65)][1 + 0.0009(Vr_{b-c} - 120)] \qquad [12.7]$$

Fig. 12.3   Major/minor junction with ghost islands and right-turning lane

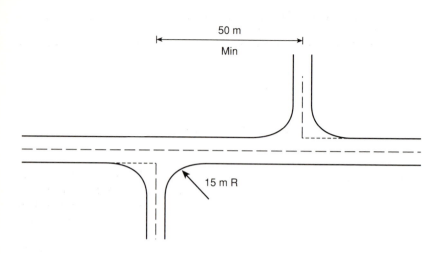

Fig. 12.4 Typical layout for staggered junction

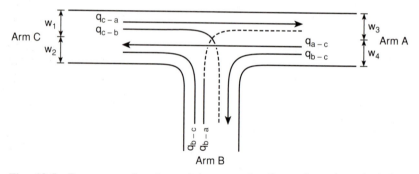

Fig. 12.5 Parameters for determining capacity flows through major/minor junction

where w_{b-a} denotes the lane width available to vehicles waiting in the stream B – A, Vr_{b-a} and Vl_{b-a} the corresponding visibilities, and so on. Capacities derived from these formulae are always shown in pcu/hr. Distances should be input in metres. A negative value indicates zero capacity.

These formulae are clearly complicated and working through them manually invites mistakes. In order to calculate capacities of streams through a range of feasible junction designs, it may be considered helpful to set these formulae on a spreadsheet (SUPERCALC or LOTUS 1-2-3), as, having once set the system up, it is possible to input alternative geometric layouts very quickly.

A specialist computer program, PICADY,[8] is available from the

Department of Transport for undertaking a complete range of analyses or traffic flows through a major/minor junction.

PICADY will also model queue lengths and delays at major/minor junctions.

Traffic signal controlled junctions

One of the commonest forms of junction control in urban areas is the traffic signal. These originally evolved from the semaphore arm type which had been introduced to ensure safety on railways in the first half of the nineteenth century, but the introduction of electrical three light systems became essential with the growth of road traffic after 1919. Most major junctions between all-purpose roads in towns and cities in developed countries are now controlled by coloured lights, which usually operate on a variation of red for stop, amber for caution and green for go.

The major feature of a signal controlled junction is that priority is not always given to the same traffic stream, but can be varied with time through the cycle of the signals. Thus, it is not necessary to have permanently defined major and minor paths, as with a major/minor priority intersection. Because priority is defined by a light targetted directly at approaching traffic, visibility requirements are not as rigid as with other forms of layout and the amount of space taken up by a signal controlled junction is substantially less than for a roundabout dealing with the same amount of traffic.

The application of modern computer technology has allowed additional advantages to be gained through linking a series of signals to a central computer. A wide range of systems is possible, from a simple linked system arranged to minimise delay through a sequence of two or more junctions to a full scale Urban Traffic Control (UTC) system linking a hundred or more intersections over a wide area. With the increase in interest in on-street running of rapid transit vehicles, such as those now operational in Manchester and Hong Kong, that under construction in Sheffield and those planned in several other centres, research is in hand to link the computer contolled traffic signals for road vehicles on urban streets and those on fixed rail tracks.

As with major/minor junctions and roundabouts, the basis of the design of a signal controlled junction is the relationship between some measure of the demand (DRF) on each flow path through the junction and the capacity of that flow path. However, there is another variable to be introduced: the setting of the signal. Thus, capacity is developed from saturation flow, which is defined as the maximum amount of traffic, in pcu/hr, that could cross the stop line assuming the light was continuously green and there was a continuous stream of traffic on the approach. The 'capacity' of the approach is related to the saturation flow by the proportion of the time that the light is green.

In the United Kingdom, traffic signals operate on a four period cycle: red, red and amber, green, amber and back to red again. The red and amber period is of standard two second duration and the amber period of standard three seconds. The designer of the installation has to determine the durations of the red and green periods based on his knowledge of demand and capacity. Similar, but not identical, arrangements apply in other countries. In the UK, it is an offence to pass a traffic signal at red. Offending drivers can be prosecuted, fined and have penalty points awarded against them which can, cumulatively, lead to a driving ban.

The design guides for traffic signals are based on work done at the Transport and Road Research Laboratory in the 1950s, and originally published in 1958.[9,10] The development of the procedures can be followed through the standard texts on the subject.[11,12,13]

Simple traffic signal installations controlling a single, isolated junction, will normally operate on the basis of a fixed time sequence. The switching device may be an electro-mechanical switch or, on modern installations, a programmable microchip, and it is likely to be located in a pillar or box located close to the junction.

It is likely to be able to respond to five different conditions which, apart from the last one, are time or traffic flow dependent:

1. Morning peak
2. Evening peak
3. Off peak
4. Very low traffic at night
5. Manual override

Programs (1), (2) and (3) are determined by time of day and the settings will always change from, say, (1) to (3) at, say, 9.30 a.m. The way that this works is that programs (1), (2) and (3) determine the *maximum* green times for a given phase of the signals. Thus, if the traffic flow ceases, the phase will terminate early. With low flows on all approaches, all the phases will terminate as soon as traffic demand has been met, which is one version of program (4). There will also be a *minimum* green time associated with a phase, which will be at least seven seconds, to ensure that all vehicles can clear the junction in safety before another flow is released.

It may be that, overnight, all the approaches are normally at red until a vehicle approaches, at which point the light on that approach turns green for a fixed time period and then returns to red. Variations on this theme will allow for the light to remain green until a vehicle approaches on an alternative leg, or for one approach to be given a green light unless a vehicle is detected elsewhere.

Traffic signal junctions will always be equipped with vehicle detectors, which may be an induction loop buried beneath the surface,[14,15] a rubber pressure pad built into the surface or microwave detector mounted on the signal head.[16]

By far the most common is the induction loop buried 25–50 mm beneath the surface of the carriageway at set intervals before the stop line. A small electrical charge is passed through the loop, setting up a magnetic field. When the field is penetrated by a magnetic object, a vehicle, then its presence is detected. Details of installation are given in Department of Transport Advice Notes TA13/81 and TA14/81.[17,18]

Program (5) allows for the police requirement that they must always be able to take over manual control of the lights in the event of an emergency.

Each program will determine the length of time allocated to each phase of the signal cycle. This is illustrated in Figure 12.6, which also defines the relationship between a signal phase and a signal stage.

Measurement of demand is not dependent upon the type of junction and the same considerations as in the case of a major/minor junction apply. The designer must arrive at a sensible conclusion as regards the DRF for which the installation is to be designed. It is worth remembering, however, that quite short term considerations might be appropriate for the design of the signals, as altering the settings is a simple and low cost operation. Indeed, with a traffic responsive UTC system, the settings are amended with every cycle of the signals in order to maximise the benefits for whatever traffic conditions apply.

As with other forms of junction, demand is represented by a traffic flow, measured in pcu/hr. The values to be applied to convert vehicles to pcus are dependent on the manoeuvres being undertaken by the traffic at the time, and pcu values are therefore different for traffic signals from those used for major/minor junctions and roundabouts. Values determined experimentally by the Transport and Road Research Laboratory in 1986[19] are as shown in Table 12.1

The designer must start by determining the layout of the intersection, or a range of possible layouts. As with other types of junction, the constraints will depend entirely upon the particular situation being considered and it is the designer's task to identify a 'best' solution from the infinite range of possibilities that must always exist. This, then, is a task to be undertaken using calculator and drawing board or computer-aided design system. The eventual output must be a layout, and associated signal settings, which satisfies the demand or, if this is not feasible, maximises safety and minimises delay.

Extensive guidance on suitable layouts for signal controlled junctions is given in a range of Department of Transport Technical Memoranda[20,21]. A typical layout, together with appropriate staging diagrams, is shown in Figure 12.7

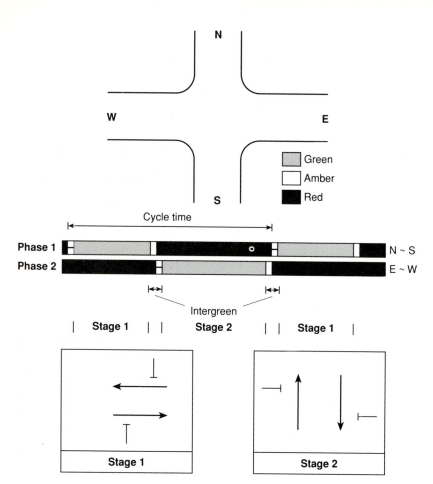

Fig. 12.6   Traffic signal phases

Table 12.1   Pcu values for use at traffic signal installations (Source: Reference 19)

Light vehicles	1.0
Medium commercial vehicles	1.5
Heavy commercial vehicles	2.3
Buses and coaches	2.0
Motor cycles	0.4
Pedal cycles	0.2

Definition of categories:
Light vehicles: 3 or 4 wheeled vehicles
Medium commerical vehicles: Vehicles with 2 axles but more than 4 wheels
Heavy commercial vehicles: Vehicles with more than 2 axles

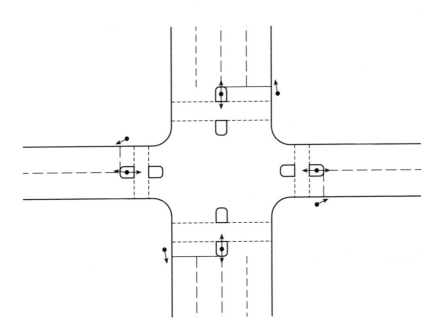

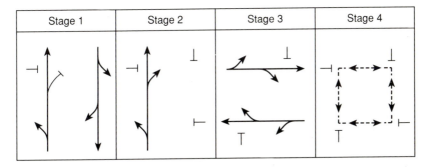

Fig. 12.7 Layout and staging diagram for signal controlled junction

Having determined the layout, it is possible to determine the saturation flow for each path through the junction using a series of predictive equations developed by the Transport and Road Research Laboratory and shown in reference 19.

The factors affecting saturation flow have been found to be:

Number of lanes
Width of lanes
Position of lanes
Turning movements
Approach gradients
Weather conditions

Looking at these in more detail, we can make the following comments.

Number of lanes

Saturation flow will increase the more lanes of traffic that there are using the approach. It has been found that there is not much interaction between the lanes and that the total saturation flow of the approach is equal to the sum of the saturation flows of each lane measured individually.

Thus

$$S_A = \sum_{i=1}^{n} S_i \qquad [12.8]$$

where S_A = Saturation flow of the approach; S_i = Saturation flow of Lane i; n = Number of lanes.

Width and position of lanes

The average width of a lane on the approach to a traffic signal is 3.25 m., and by observation of non-nearside lanes of this width at different sites, the mean value of saturation flow has been found to be 2080 pcu/hr. For nearside lanes, it is 1940 pcu/hr.

Where traffic was all going straight ahead, these figures increased by 100 pcu/hr/m for increased lane width.

Turning movements

For individual lanes containing *unopposed* turning traffic, saturation flow decreases for higher proportions of turning traffic and lower radii of turn.

Where turning traffic is *opposed* by traffic approaching from a different direction, the saturation flow depends heavily upon the number of gaps in the opposing flow, the amount of storage space for right-turners to await an opportunity to move without interfering with the progress of straight ahead traffic, and the frequency of change of phase in the cycle.

We will thus use two predictive equations, one for where there are no opposed turning movements and one for where there are opposed turning movements.

Note that it is not the turning movements themselves that are critical: it is whether there is any opposing flow.

Gradients

Saturation flow is decreased on uphill gradients by 2% for each 1% of gradient measured over a distance of 60 m from the stop line. It is unaffected by downhill gradients.

Weather conditions

Saturation flow is about 6% lower when the road is wet compared to when it is dry.

The predictive equations determined from a large number of traffic movements through observed signal controlled junctions are as follows:

(a) *Unopposed streams in individual lanes*

$$S_1 = (S_0 - 140\delta_n)/(1 + 1.5f/r) \qquad [12.9]$$

where

$$S_0 = 2080 - 42\delta_G G + 100(w_1 - 3.25) \qquad [12.10]$$

(b) *Streams containing opposed right turning traffic in individual lanes*

$$S_2 = S_g + S_c \qquad [12.11]$$

where

$$S_g = (S_0 - 230)/[1 + (T - 1)f] \qquad [12.12]$$

and

$$T = 1 + 1.5/r + t_1/t_2 \qquad [12.13]$$

$$t_1 = 12X_0^2/[1 + 0.6(1 - f)N_s] \qquad [12.14]$$

$$t_2 = 1 - (fX_0)^2 \qquad [12.15]$$

and

$$S_c = P(1 + N_s)(fX_0)^{0.2}3600/\lambda_c \qquad [12.16]$$

$$X_0 = q_0/n_1 S_0 \qquad [12.17]$$

The notation in the above equations is as follows:

S = Saturation flow (pcu/hr) with subscript to differentiate between lanes or types of lane

δ_G, δ_n = dummy variables for gradient/lane position (=0 for downhill, non-nearside lane or =1 for uphill, nearside lane)

f = proportion of turning traffic in a lane on a scale from 0 to 1

r   = radius of vehicle paths (m)
G   = gradient (%)
w_1 = lane width (m)
X_0 = degree of saturation on the opposing arm (i.e. the ratio of the actual flow on the opposing arm to the saturation flow on that arm)
N_s = number of storage spaces for right-turners
λ = proportion of cycle time that is effectively green for phase being considered on a scale from 0 to 1
c   = cycle time (secs)

Using these equations, it is possible to determine the saturation flow for any lane on any approach to a junction, the geometrical properties of which are known. For opposed flows, it is also necessary to know or to assume a value for the opposing traffic stream.

The following examples[22] show how these equations may be applied. Consider the layout of a T-junction shown in Figure 12.8. In this situation, the turning movements are *unopposed*.

Assume that this is going to be a two stage installation, with staging diagram as shown in Figure 12.9

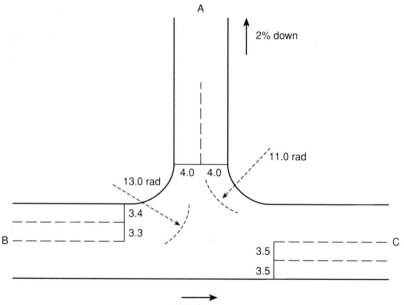

Fig. 12.8   Layout of signal controlled junction used in example

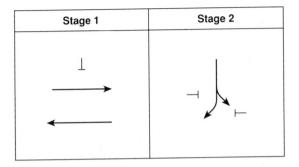

Stage 1	Stage 2

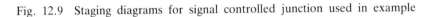

Fig. 12.9 Staging diagrams for signal controlled junction used in example

Approach A is a one-way street for traffic turning into the junction, so traffic on both Approach B and Approach C must go straight ahead. There are no turning movements, either opposed or otherwise.

Traffic on Approach A, however, must turn, for there is no road straight ahead, but the turning movements are unopposed.

So equations 12.8, 12.9 and 12.10 apply in this case.

Consider Approach B. There are two lanes, so the saturation flow of the approach is the sum of the saturation flows for each lane (equation 12.8).

First, we need to find the saturation flow of the nearside lane (S_n)

$$S_0 = 2080 - 42\delta_G G + 100(w_1 - 3.25) \qquad [12.10]$$

The values we have to substitute are

$$\delta_G = 0 \text{ for downhill gradient}$$

G is therefore irrelevant

$$w_1 = 3.4 \text{ (stop line width in m.)}$$

so

$$S_0 = 2080 + 100(0.15) = 2095$$

and

$$S_1 = (S_0 - 140\delta_n)/(1 + 1.5f/r) \qquad [12.9]$$

The values to substitute here are

$$S_0 = 2095 \text{ (from previous calculation)}$$

$$\delta_n = 1 \text{ (for nearside lane)}$$

$$f = 0 \text{ (proportion of vehicles turning)}$$

therefore r does not apply so

$$S_n = S_1 = (2095 - 140)/(1 + 0)$$

$$= 1955 \text{ pcu/hr}$$

We can now do exactly the same calculation for the offside lane, bearing in mind that the values of w_1 and δ_n are going to be different.

$$S_0 = 2080 + 100 (3.2 - 2.5)$$

$$= 2075 \text{ pcu/hr}$$

$$S_f = S_1 = (2075 - 0)/(1 - 0)$$

$$= 2075 \text{ pcu/hr}$$

The total saturation flow of the whole approach is

$$S_B = S_n + S_f$$

So, consider Approach A.

The turning movements are still unopposed, so we are still using equations 12.8, 12.9 and 12.10. But this time we will have to take the parameters f and r into account, as follows:

f is the proportion of turning traffic measured on a scale of 0 to 1. On approach A there is provision for traffic to go straight ahead so all the traffic is turning traffic, i.e. f=1

r is the average turning radius measured in metres. This is the function of the geometry of the junction and is shown on Figure 12.8.

So, for the left hand, nearside lane

$$S_0 = 2080 - 42\delta_G \; G + 100(W_1 - 3.25) \qquad [12.10]$$

$$\delta_G = 1 \text{ for uphill gradient}$$

$$G = 2$$

$$w_1 = 4.0$$

$$S_0 = 2080 - 42 \times 1 \times 2 + 100(4.0 - 3.25)$$

$$= 2071$$

$$S_n = (S_0 - 140 \; \delta_n)/(1 + 1.5f/r) \qquad [12.9]$$

$$S_0 = 2071$$

$$\delta_n = 1$$

$$f = 1$$

$$r = 11$$

$$S_n = (2071 - 140)/(1 + 1.5 \times 1/11)$$

$$= 1699 \text{ pcu/hr}$$

Example of saturation flow involving opposed turning movements

The next step is to consider the situation in which there are opposed turning movements.

This situation occurs at what is probably the most common form of signal controlled junction; the straightforward crossroads. In fact, cross-roads are discouraged for reasons of accident risk, unless there is signal control (or a roundabout). An example layout is shown in Figure 12.10 and design flows in Figure 12.12.

The junction shown is controlled by a simple 2-stage signal, one stage showing green to A and C and the other showing green to B and D.

Because the (right) turning movements are opposed, we are going to use equations 12.11–12.17 and equation 12.9.

Saturation flow in this case is heavily dependent upon the behaviour of right-turning vehicles, and this in turn will depend upon the traffic conditions in the opposing flow.

If the opposing flow is small, a right-turner is likely to be able to

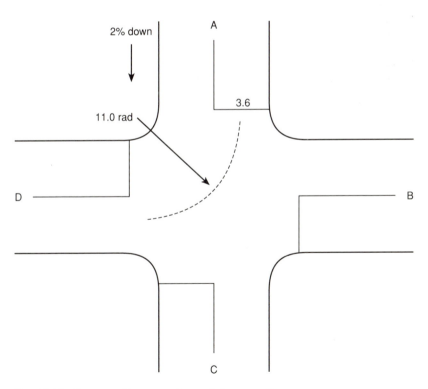

Fig. 12.10 Layout of crossroads used in example

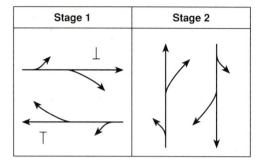

Fig. 12.11 Staging diagram used in crossroads example

move immediately, and no queue will form. However, if the opposing flow is large, a right-turner has to wait until a suitable gap appears. This may not happen until the next change of phase, but it should happen then. If there are large numbers of right turning vehicles, a queue will form and this queue may impede the progress of non-turning traffic.

From this statement, it is possible to see the parameters which must be considered if we are to model a situation involving opposed turning traffic, in addition to the parameters that we have already used. They are:

- the amount of traffic on the opposing arm, measured in terms of the ratio of actual flow: saturation flow. (X_0 in equations 12.15, 12.16 and 12.17)
- the number of storage spaces available for right-turning vehicles. (N_s in equations 12.14 and 12.16)
- proportion of the cycle that is effectively green for the phase under consideration. (λ in equations 12.16 and 12.17).

This means that we are going to have to make assumptions about cycle times and opposing flows if we are to carry out the calculations.

So, as a starting point, let us make some assumptions:

degree of saturation of opposing flow, $X_0 = 0.7$
number of storage spaces, $N_s = 3$
pcu factor, $P = 1.15$
cycle time $= 60$ secs
effective green time for phase A/C $= 32$ secs
proportion of turning traffic $= 0.2$.

We can now determine a value for saturation flow in this lane (S_2 in equation 12.11)

$$S_2 = S_g + S_c \qquad [12.11]$$

$$S_c = P(1 + N_s)(fX_0)^{0.2}3600/(\lambda \times c) \qquad [12.16]$$

where P = pcu factor = 1.15; N_s = Number of storage spaces = 3; f = Proportion of turning traffic = 0.2; X_0 = Degree of saturation of opposing flow = 0.7; λ = Proportion of cycle which is effectively green = 32/60 = 0.53; c = Cycle time (in seconds).
So

$$S_c = 1.15 \times (1 + 3) \times (0.2 \times 0.7)^{0.2} \times 3600/(0.53 \times 60)$$

$$= 351.5 \text{ pcu/hr}$$

$$S_g = (S_0 - 230)/[1 + (T - 1)f] \qquad [12.12]$$

$$S_0 = 2080 - 42\delta_G + 100(w_1 - 3.25)$$

The approach is downhill, so $\delta_G = 0$

$$S_0 = 2080 + 100(3.6 - 3.25)$$

$$= 2115 \text{ pc/hr}$$

$$t_1 = 12X_0^2/[1 + 0.6(1 - f)N_s] \qquad [12.14]$$

$$= 12 \times (0.7)^2/[1 + 0.6 \times (1 - 0.2) \times 3$$

$$= 2.410$$

$$t_2 = 1 - (fX_o)^{0.2} \qquad [12.15]$$

$$= 1 - (0.2 \times 0.7)^2$$

$$= 0.980$$

$$T = 1 + 1.5/r + t_1/t_2 \qquad [12.13]$$

where r = radius of turning movement = 11.0; T = 3.595.
Going back to equation 12.12

$$S_g = (S_0 - 230)/[1 + (T - 1)f]$$

$$= (2115 - 230)/(1 + 2.594 \times 0.2)$$

$$= 1241.1 \text{ pcu/hr}$$

Saturation flow for this lane

$$S_2 = S_g + S_c \qquad [12.11]$$

$$= 1241.1 + 351.5$$

$$= 1592.6 \text{ pcu/hr}$$

If we had not taken turning traffic into account at all, we could have done a calculation based on Equation 3 with f = 0 and obtained a value of 1975 pcu/hr.

If we had used Equation 12.10 and put f = 0.2 (i.e. we took account of turning traffic but assumed it was unopposed) saturatio n flow would have been 1922 pcu/hr. So there is a substantial drop in saturation flow when turning movements are opposed.

Having determined the saturation flow for each approach, the next stage in the determination of the signal settings is to determine the critical y-value for each stage.

The y-value on each approach is the ratio of the design flow to the saturation flow. Consider the flows and layout shown in Figures 12.10, 12.11 and 12.12 and the saturation flows to be

$$S_A = 1465$$

$$S_B = 1040$$

$$S_C = 1185$$

$$S_D = 1420$$

Assuming that the signal installation is a 2-stage signal with A and C green together and B and D green together, then

$$y_A = 622/1465 = 0.42$$

$$y_B = 327/1040 = 0.31$$

$$y_C = 466/1185 = 0.39$$

$$y_D = 611/1420 = 0.43$$

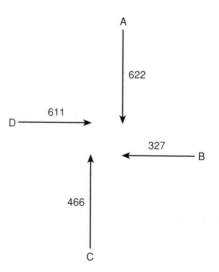

Fig. 12.12   Design flows for crossroads example

We can now select the critical y-value. It is the larger of all the y_X values that form a particular stage, i.e. that are green together. Thus

$$y_I = 0.42 \text{ (the larger of } y_A \text{ and } y_C)$$

and

$$y_{II} = 0.43 \text{ (the larger of } y_B \text{ and } y_D)$$

We can also say that $Y = y_I + y_{II} = 0.85$. This is the equivalent of RFC ratio in major/minor junctions where it was necessary to keep it greater than 0.85 to ensure that the junction was working without undue delay being caused. The equivalent figure for traffic signals is 0.90.

When traffic moves off from a traffic light, the flow across the stop line builds up gradually to the point where the saturation flow is reached. Flow then becomes steady until the light turns amber at the beginning of the next cycle. Some drivers will manage to stop almost immediately whereas others will be rather tardy about noticing the signals. Thus, on average, the flow begins to fall.

This is shown graphically in Figure 12.13

From Figure 12.13 it is possible to see what is meant by saturation

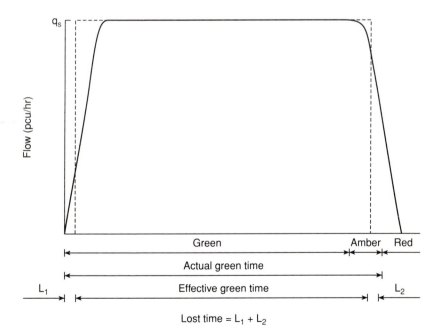

Fig. 12.13 Definition of lost time and effective green time

flow and the relationship between actual green time (the period of time during which the light shows either green or amber) and the effective green time, which is the length of time which would be required to get a given flow over the stop line if the flow started and stopped simultaneously on the change of light colour on the signals. Thus, effective green time is the length of time during which saturation flow would have to be maintained to get the same amount of traffic through the light as does in fact pass during a real green period. It is represented by the rectangle on Figure 12.13 which has the same area as that under the actual flow curve.

Effective green time is less than actual green time, and the difference between the two is called lost time.

There is always an element of lost time every time the lights change, because there must be a safety period between the time at which one traffic stream is flowing through the junction and the time at which a conflicting stream is using the same area of road space. At minimum, this lost time is regarded as being two seconds per change of phase but it is frequently longer if, for instance, the distance between the stop line and the conflict point in the centre of the junction is greater than average.

Webster developed a series of formulae to determine the optimum signal settings so as to minimise total delay on all streams approaching the junction. His formula for optimum cycle time is

$$c_o = (1.5L + 5)(1 - Y)^{-1} \qquad [12.18]$$

where c_o = optimum cycle time; L = total lost time per cycle; $Y = y_1 + y_2 + y_3 + \ldots + y_n$.

All time values are measured in seconds. A maximum value of c_o is normally taken as 120 seconds, as longer periods of red lead to driver apprehension and hence to accidents.

Thus, from the calculation of saturation flow for each approach to the signal, it is possible to determine the optimum signal setting for any given traffic conditions.

Modern control equipment, which opens up the possibility of

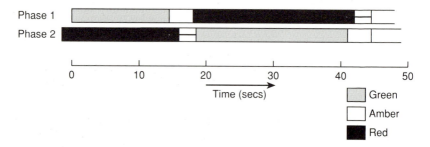

Fig. 12.14   Settings of a traffic signal resulting from calculation in Box 12.1

programming a wide variety of traffic conditions into a single controller can make a number of other options available to the designer. Turning traffic can be handled using late release or early cut off. For instance, in our example, we might want to cope with a large number of right-turners approaching on A. We might introduce an extra stage which allowed traffic to turn right from A whilst traffic approaching on C was stopped. However, we can achieve the desired effect without adding an extra stage. We can use a late release or early cut-off. That is, the settings are the same as we worked out before, but instead of A and C turning to green at the same time, C goes green after A. Thus, for a short time at the beginning of each green phase on A, traffic on C is stopped and does not impede traffic wishing to turn right.

Roundabouts

There are two basic forms of roundabout available in current British practice: a normal roundabout and a mini-roundabout. From time to time, two mini-roundabouts are located so close to each other as to form a double roundabout. There is one example (in Hemel Hempstead, Hertfordshire) of a 'ring junction' which consists of a series of mini-roundabouts located around the ring of a large roundabout. Two way traffic passes around the large ring. Traffic signals are frequently found installed on roundabouts.

Roundabouts perform two functions:

1. They define priority between traffic streams at a junction, nearly always on the basis of 'priority to the right', i.e. traffic wishing to join the circulatory flow has to give way to traffic already circulating;
2. They cause traffic to divert from a straight line path, meaning that it must slow down to pass through the junction.

Normal roundabouts are those having a central island greater than 4 m in diameter. They are suitable for use in the following situations:

1. At the end of a section of new road where it is desirable to slow traffic or cause it to change direction.
2. At a point where road standards change from rural to urban.
3. As part of the road layout for a new industrial or commercial estate.

Normal roundabouts are relatively expensive in terms of land utilisation. They are designed on the basis of entry capacity making use of flared approaches. The earlier basis of roundabout design, using weaving lengths, is not used.

Mini-roundabouts are those having a central island of less than 4 m diameter, and frequently no more than 2 m. The island may be no more substantial than white road markings. They are used to improve the performance of an existing road junction where all approaches are subject to a 30 mph speed limit.

Typical layouts for standard and mini-roundabouts are shown at Figures 12.15 and 12.16.

The operation of a roundabout depends upon sufficient gaps occurring in the circulating flow to allow traffic on the arms of the roundabout to enter it. It therefore has certain similarities with the situation at a major/minor junction, and the design procedure is comparable. The situation is more complicated because there is no single, definable major road flow for which the junction can be designed. The circulating flow will itself depend upon the operation of the junction. If we consider a four-legged roundabout, having entries defined as A, B, C and D, and circulating flows defined as q_{ab}, q_{bc}, q_{cd} and q_{da} (see Figure 12.17) it is clear that q_{ab} will depend upon q_{da}

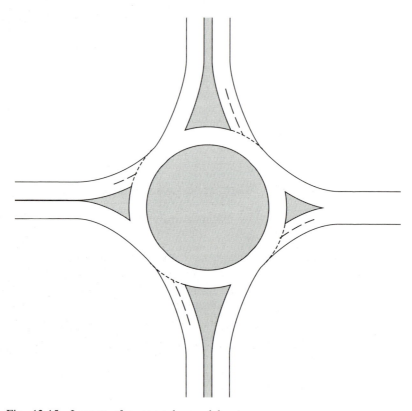

Fig. 12.15 Layout of a normal roundabout

(a) Splayed approach to normal roundabout on single carriageway trunk road (A5 Oswestry By-pass, Shropshire)

(b) Mini roundabout at junction of distributor road with principal radial route (A484 Gorseinon, West Glamorgan)

Fig. 12.16

(c) Mini roundabout at access to industrial premises (Gorseinon, West Glamorgan)

(d) Normal roundabout at junction of by-pass and existing distributor road (Gowerton, West Glamorgan)

Fig. 12.16 continued

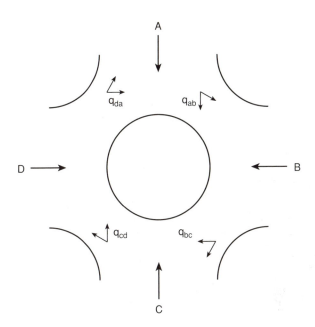

Fig. 12.17   Parameters for determining flow on a roundabout

which will depend upon q_{cd} which will depend upon q_{bc} which will depend upon q_{ab}, which is where we started. Determination of capacity flows on a roundabout is therefore an iterative procedure, as illustrated in Figure 12.18.

The procedure for determining capacity of roundabouts is laid down in a series of technical memoranda published by the Department of Transport.[5,23,24] As with major/minor junctions, it is first necessary to determine a Design Reference Flow based on forecast traffic levels. This must be expressed in pcu/hr, but the weighting factor adopted is different from other junctions, so, in this case, cars and light vans have a pcu value of 1.0 and heavy vehicles a pcu value of 2.0.

The predictive equations for the capacity of an entry to a roundabout are based on observations of a large number of sites reported by Kimber[25] and depend upon the geometry of the entry and the circulating flow. The geometric characteristics found to have a significant effect on capacity were:

1. Entry width (e)
2. Approach half-width (v)
3. The average effective length (l) over which the flare is developed
4. The sharpness of the flare $S = 1.6(e - v)/l$
5. The entry radius (r)
6. The angle of entry (ϕ)
7. The inscribed circle diameter (D).

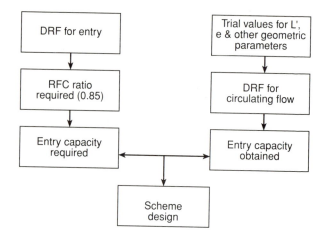

Fig. 12.18 Procedure for roundabout design

These parameters are defined in Figure 12.19.
The predictive equations for saturation flow Q_s at entry given a circulation flow Q_c are as follows:

$$Q_s = k(F - f_c Q_c) \text{ when } f_c Q_c < F \qquad [12.19]$$

$$Q_s = 0 \qquad \text{when } f_c Q_c > F \qquad [12.20]$$

where $k = 1 - 0.00347(\phi - 30) - 0.978(1/r - 0.05)$; $F = 303x_2$; $f_c = 0.21t_D(1 + 0.2x_2)$; $t_D = 1. 05/(1 + M)$; $M = e^j$; $j = (D - 60)/10$.

In these equations, ϕ is measured in degrees and all other dimensions in metres.

Working through an iterative process of this nature is clearly a lengthy task. The Department of Transport has available a computer program, ARCADY,[26] for undertaking this work.

Grade separated interchanges

Grade separation will be provided to ensure that there is no at-grade conflict between intersecting traffic streams in the following circumstances:

1. Where one or more of the roads carrying through traffic is classified as being a motorway.
2. Where predicted traffic flow on the major route is sufficiently high to mean that design of an at-grade junction would use up so much space as to render it unfeasible or uneconomic.

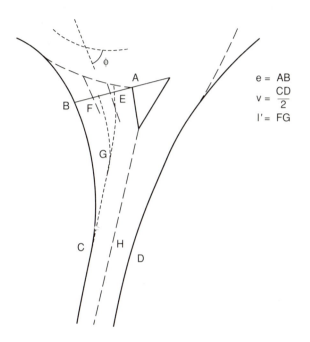

$$e = AB$$
$$v = \frac{CD}{2}$$
$$l' = FG$$

Fig. 12.19 Geometric parameters for roundabout design

3. Where provision of grade separation makes sensible use of the available land and levels.
4. Where traffic on the major route is likely to be travelling at high speed and safety considerations discriminate against an at-grade intersection.

Because of the need for bridges and possibly other structures, and because of the amount of land used up, grade separation will undoubtedly be a more expensive option. Such a junction is likely to be individually designed, but it is possible to identify a number of standard layouts. These are illustrated in Figures 12.20 and 12.21. Further details are provided in the relevant Department of Transport documents.[27]

Box 12.1:

Example of determination of traffic signal settings:

Consider the simple crossroads defined in Figure 12.12. The saturation flow on each approach is as follows:

Box 12.1 continued:

A 1876 pcu/hr
B 1988 pcu/hr
C 2012 pcu/hr
D 1777 pcu/hr

A traffic survey conducted during the morning peak has established the following flows on each approach to the junction:

A 625 pcu/hr
B 819 pcu/hr
C 212 pcu/hr
D 747 pcu/hr

Step 1: Determine the y-values on each approach by use of

$$y = \text{actual flow (q)/saturation flow (s)}$$

$$y_A = 625/1876 = 0.33$$

$$y_B = 819/1988 = 0.41$$

$$y_C = 212/2012 = 0.11$$

$$y_D = 747/1777 = 0.42$$

Step 2: Select critical y values for each stream (assuming two stage signal with light green on A-C or B-D) and hence determine Y-value.
Critical values are y_A and y_D, hence

$$Y = y_A + y_D = 0.33 + 0.42$$
$$= 0.75$$

Step 3: Determine optimum cycle time from Webster's formula (eqn. 12.18)

$$c_o = (1.5L + 5)/(1 - Y)$$

We know that the minimum lost time is 2 seconds per change of phase. This is a two-phase system so the minimum value of L is 4 seconds. It may be that the layout of the junction would lead us to increase this value, but for our purpose here, we will use L=4, hence

$$c_o = (1.5 \times 4 + 5)/(1 - 0.75)$$
$$= 44 \text{ seconds}$$

Step 4: Allocate the effective green time between the two phases. We know that the total effective green time is 40 seconds, the cycle time less the lost time ($c_o - L$).

We divide this between the phases in proportion to the critical y-values.

Thus the A-C phase will get (0.33/0.75) × 40 seconds of effective green time = 17.6 seconds and the B-D phase will get the remainder = 22.4 seconds.

These effective green times then have to be converted to actual signal settings by allocating the lost time that we have. This will give signal settings which are shown in Figure 12.14.

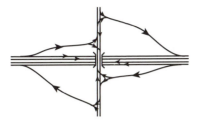
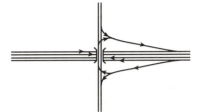

(*a*) Diamond

(*b*) Half-diamond

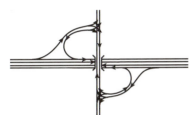
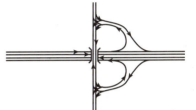

(*c*) Half-Cloverleaf
Quadrants 1 and 3

(*d*) Half-Cloverleaf
Quadrants 2 and 3

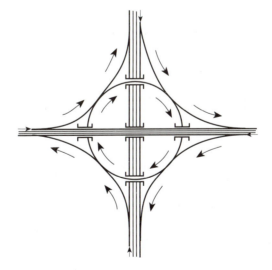

(*e*) 3 Level Roundabout

Fig. 12.20   Grade separated junction layouts (Source TA48/86)

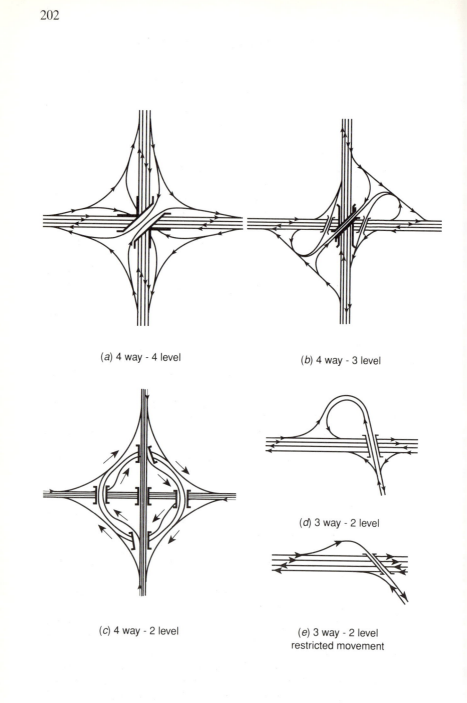

(a) 4 way - 4 level

(b) 4 way - 3 level

(d) 3 way - 2 level

(c) 4 way - 2 level

(e) 3 way - 2 level
restricted movement

Fig. 12.21 Grade separated interchange layouts (Source TA48/86)

References and further reading — Chapter 12

1. Department of Transport. *Traffic Signs Manual: Chapter 5. Road markings.* HMSO. London. 1985.
2. Department of Transport. *Traffic Signs Manual: Chapter 1. Introduction.* HMSO. London. 1981.
3. Department of Transport. *Traffic Signs Manual: Chapter 3 Regulatory signs.* HMSO. London. 1986.
4. Department of Transport. *Traffic Signs Manual: Chapter 4. Warning signs.* HMSO. London. 1986.
5. Department of Transport. Departmental Advice Note TA23/81. *Determination of Size of Roundabouts and Major/Minor Junctions.* Department of Transport. London. 1981.
6. Department of Transport. *Traffic Appraisal Manual.* Section 12. Department of Transport. London. 1991.
7. Department of Transport. Departmental Advice Note 20/81. *The Layout of Major/Minor Junctions.* Department of Transport. London. 1981.
8. Siemans M C. *PICADY2 – an enhanced program for modelling traffic capacities, queues and delays at major/minor junctions.* TRRL Report RR36. Transport and Road Research Laboratory. Crowthorne. 1985.
9. Webster F V. *Traffic Signal Settings.* Road Research Technical Paper No. 39. London. HMSO. 1958.
10. Webster F V and Cobbe B M. *Traffic Signals.* Road Research Technical Paper No. 56. London. HMSO. 1966.
11. O'Flaherty C A. *Highways: Volume 1. Traffic Engineering and Planning.* 3rd Edition. London. Edward Arnold. 1986.
12. Salter R J. *Traffic Analysis and Design.* 2nd Edition. London. Macmillan. 1983.
13. Salter R J. *Traffic Engineering – Worked examples and Problems.* London. Macmillan. 1981.
14. Department of Transport. Departmental specification MC0100. *Induction loop vehicle detecting equipment.*
15. Department of Transport. Departmental specification MC0108. *Siting of inductive loops for vehicle detecting equipment at permanent road traffic installations.*
16. Department of Transport. Departmental specification MC0114. *Microwave vehicle developing equipment.*
17. Department of Transport. Technical Memorandum TA13/81: *Requirements for the installation of traffic signals and associated equipment.* Reprinted March 1990. DTp. London. 1981.
18. Department of Transport. Technical memorandum TA14/81. *Procedures for the installation of traffic signals and associated control equipment.* Reprinted March 1990. DTp. London. 1981.
19. Kimber R M, McDonald M and Housell N B. *The prediction of*

saturation flows for road junctions controlled by traffic signals. TRRL Research Report No 67. Transport and Road Research Laboratory. Crowthorne. 1986.

20. Department of Transport Technical Memorandum TA18/81. *Junction layout for control by traffic signals.* Reprinted February 1989. Department of Transport. London. 1989.

21. Department of Transport Technical Memorandum TA15/81. *Pedestrian facilities at traffic signal installations.* Reprinted March 1990. Department of Transport. London. 1990.

22. Macpherson G. *Stop, go, saturation flow: a learning exercise for civil engineers.* Sheffield City Polytechnic. 1987.

23. Department of Transport. Departmental standard TD16/84: *The geometric design of roundabouts.* DTp. London. 1984.

24. Department of Transport. Departmental advice note TA42/84: *The geometric design of roundabouts.* DTp. London. 1984.

25. Kimber R M. *The traffic capacity of roundabouts.* TRRL Report LR942. Transport and Road Research Laboratory. Crowthorne. 1980.

26. Semmens M C. ARCADY2: *An enhanced program to model capacities, queues and delays at roundabouts.* TRRL Research Report 35. Transport and Road Research Laboratory. Crowthorne. 1985.

27. Department of Transport. Technical memorandum TA48/86. *Layout of grade separated junctions.* DTp. London. 1986.

Chapter 13

Urban Traffic Control

Linked signals

In Chapter 12, we have looked at the way in which traffic signals can be used to control traffic flow at isolated junctions. Their purpose is threefold:

1. to separate conflicts in time and hence contribute to road safety;
2. to distribute the time available between traffic streams in such a way as to minimise the total delay to all traffic;
3. to control the formation of traffic queues in such a way as to minimise the obstruction caused by standing vehicles to other traffic.

Modern technology, and particularly the availability of low cost, powerful computers, opens up other possibilities by controlling the way in which different signals interact.

The most simple form of 'urban traffic control' is the linking of signals along a principal route to a single controller. A common situation faced by traffic travelling along a radial route into or out of a city centre is to be faced with a succession of signal controlled junctions where feeder routes join the main road. A part of such a situation is illustrated in Figure 13.1.

Consider traffic travelling in the direction A – B – C, where each of

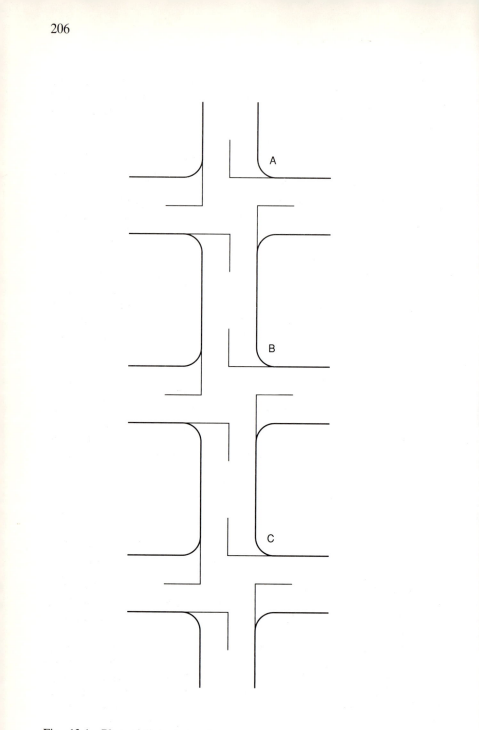

Fig. 13.1 Plan of linked signal system

the junctions is signal controlled. When the light at A turns green, traffic will move off towards B. If the settings have been determined without reference to each other, when traffic arrives at B it might find that the light is green, or it might not. If it is green, it will simply proceed towards C. But, unless it happens to have turned green at the moment that the first vehicle from A arrives, some of the green time will have been wasted. The light at B has been green, but no traffic is wanting to pass it, because it has been held up by the red light at A.

An alternative situation is that A and B are both green, but C is red. A queue therefore forms back along the road from C until it reaches B. At that point, traffic arriving from A cannot pass B because the way is obstructed by traffic queueing from C. Traffic wanting to cross or join the radial route at B will also be obstructed by a queue across the junction.

The solution to the problem is to co-ordinate the light settings along the primary route, by offsetting the cycle times. Thus, the light at B is timed to turn green just as the first vehicle released from A reaches it. Similarly, the light at C is timed to go green after a sufficient time interval has elapsed since B turned green to ensure that vehicles pass through C without stopping.

This system allows some control to be exercised over the formation of queues. Provided that C is timed precisely to release traffic arriving from B without stopping it, traffic travelling straight ahead on the radial route will never be delayed at C. Any queue will be formed from traffic that has come from the side road, joining the main road at B. This may be helpful if the section of road between B and C is an undesirable place for queues to form – because of numbers of parked cars or pedestrians, for instance.

Another advantage of this kind of linked signal is that it can control traffic speeds. The amount of time offset from one signal to the next is determined by the amount of time that a vehicle takes to travel that section of road. This, in turn, is dependent on vehicle speed. If a vehicle travels faster than was intended when the offset time was fixed, it will simply reach the next light too soon and have to stop, because it will still be red. The designer of the system may feel that an appropriate speed for the length of road in question is 35 km/hr. Regular drivers will soon learn that there is nothing to be gained by travelling faster than that, and will regulate their behaviour accordingly.

The main disadvantage is that the cycle time for each set of signals linked must be the same or they must be a precise multiple of a fixed time period. Otherwise, the cycles will get out of co-ordination. This doesn't matter too much if a single flow through all the signals is dominant, as often happens on radial routes into city centres. But a system such as this cannot cope with complex flows through whole networks.

208

A diagramatic representation of offset signal timings is shown in Figure 13.2.

Microprocessor signal control

The availability of microprocessor signal controls allows far more complicated situations to be effectively handled without going to the expense of a fully-fledged urban traffic control scheme.

We have seen in Chapter 12 that a typical local time switch housed in a pillar close to the junction being controlled can cope with five situations: morning peak, off-peak, evening peak, vehicle actuation for

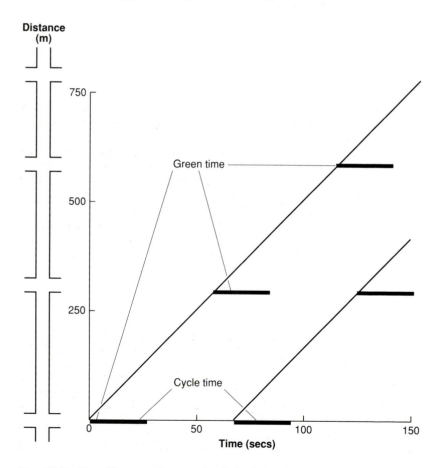

Fig. 13.2 Time/distance diagram for linked signals

light traffic flows and manual over-ride. By linking consecutive signals to the same controller we can expand this system to cover several junctions, but with the disadvantages outlined above.

We could refer to this as a three plan system. At, for instance, 7.30 in the morning, the controller will switch to Plan A (morning peak). At 9.30, it switches to Plan B (off-peak). At 4.00 in the afternoon, it goes to Plan C (evening peak), at 6.00 back to Plan B and when traffic levels fall off, vehicle actuation comes into effect. In an emergency, a police officer can come along with a key to the pillar and bring in manual control. Each plan is a different sequence of signal stages.

This system is still not very sophisticated. 'Morning peak' for instance, might have to describe a two-hour period, although there will be very significant variations in demand within that time.

What is needed is more plans, and fortunately pre-programmed microchips are able to provide them.

Local network settings

Consider the layout shown in Figure 13.3. It shows the road system passing through a Victorian out of town shopping area in the suburbs of a major industrial city.

The main radial inter-city route enters from Approach A and leaves by Approach D, which leads to the city centre. In-bound traffic is therefore confronted with three sets of signals within the space of about 250 m, with the primary flow turning right at the third signal. Out-bound traffic goes through the reverse sequence.

Approaches B and E are primary distributors of major significance within the context of the urban traffic network. B is the major route leading to a large residential area and gives access to the fringe of the

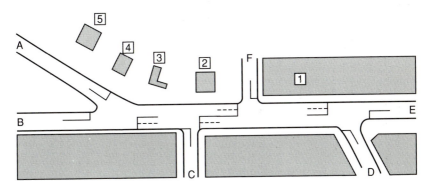

Fig. 13.3 Plan of local area network signal system

central business district, including the sites of the university and the major district hospital. Approach C is a minor route giving access to an area of 19th-century housing, mostly terraced and having low car ownership. (Less than 30% of households having one or more cars). F is a service road giving rear access to the shops and offices in Building 1 and to the rooftop car park.

Some attempt has been made to control loading and unloading to the shops in Buildings 6, 7 and 8 during peak periods. However, the nature of the businesses makes it impractical to impose any more rigid controls on the frontagers. On-street parking is prohibited.

It is clear that unco-ordinated signals would rapidly give rise to chaos as there is no space for queues to form. The most critical movements are likely to be the right-turners to D in-bound and to A out-bound, and in both cases any queue would immediately block the rest of the system.

Nor is it feasible to adopt a straightforward three-plan linked signal system as described above. The traffic movements at each of the three signal sets are quite different. For instance, in-bound traffic from Approach A, which could be expected to be substantial during the morning peak, would initially have to be merged with another substantial flow (from B). Even if C–F is considered to be a crossroads with single signal settings, demand on both C and F is likely to be intermittent. A demand actuated system would probably be installed if this junction was to be considered in isolation. The main flow then has to split again between E and D, that heading for D probably being the larger flow, which has to negotiate a right turn. The right turn into D is relatively restricted, with poor visibility and a sharp turning radius, thus reducing the saturation flow.

A feasible solution can be put forward on the basis of six stages which are illustrated in the staging diagrams shown in Figure 13.4.

There are other possible permutations and combinations of green sequences but these six will serve to illustrate the point. They give four stages, one each to give the main flows (A–D, A–E, B–E and B–D) a clear run through the area and two additional stages to cope with traffic entering and leaving by C and F.

Each stage can have a series of cycle times associated with it, which will depend upon the demand flows. These might vary within quite short timescales. We have seen above that Approach E serves a major hospital and the university. Each of these traffic generators will have distinct traffic generation patterns, which are different from the traffic generation pattern of, say, the shops in buildings 6, 7 and 8. Traffic through the area as a whole will be the result of superimposing a whole range of traffic generation patterns on top of each other, and may vary considerably over time periods as short as 5–10 minutes. As the traffic demand varies, so does the optimum of the cycle time for each of the stages 1–6.

We can help to resolve the problem by defining a number of plans.

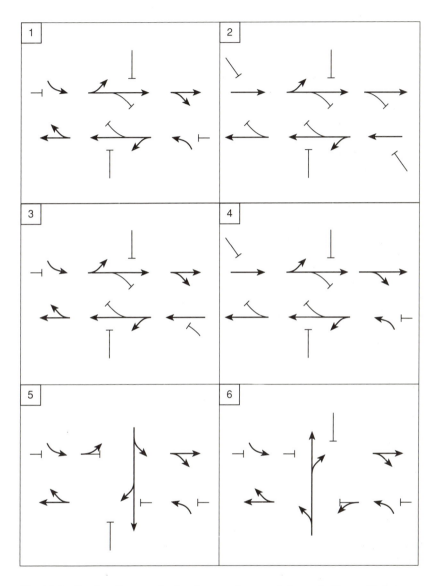

Fig. 13.4 Staging diagram for local network control shown in Figure 13.3

A plan represents a combination of each of the stages 1–6, with different timings allocated to each stage, and, hence, different cycle times. Thus, we might have something which we will call Plan 1, in which the effective green times for each stage are as follows:

Stage	Effective green time (secs)
1	35
2	25
3	15
4	18
5	0
6	0

Note that, on this plan, zero effective green time is allocated to stages 5 and 6, which are the stages which allow for traffic on the two minor approaches, C and F. We can build an element of vehicle actuation into our system by having detectors on approaches C and F which are linked to the controller. When a vehicle is identified on one of these approaches, the program will instruct the system to move into a plan which incorporates a positive green time on the approach concerned. If no vehicle has been identified, the program will miss that plan and go on to the next one.

There is no limit, other than the capacity of the controller to store data, to the number of plans that can be calculated. In practice, however, about 12 might be sufficient to cope with an area of the size illustrated.

Calculation of the demand/saturation flow ratio is a complicated operation. The demand on the second and third signals depends upon the amount of traffic released from the first and, hence, on the signal timings. We could work through the system a very large number of times, each time determining a new plan, and establishing which plan coped best with which traffic conditions. This process can be undertaken using the Department of Transport program, TRANSYT.[1,2,3,4]

TRANSYT

The objective of TRANSYT (TRAffic Network StudY Tool) is to define fixed time plans which will optimise signal settings over a network. It requires that:

1. All major junctions in the network have signals or operate on the priority rule.
2. All signals in the network have a common cycle time, or a cycle time which is an exact fraction of the common cycle time. If the area being subject to control is too large for this to be a practicable proposition then it should be divided up into a series of sub-areas, each of which will have its own plan in operation.

3. For each distinct traffic stream flowing between junctions, the flow rate, averaged over a specified time, is known and assumed constant.

Within TRANSYT, the network is represented by nodes and links. Each distinct traffic stream leading to a node is a link. A link may represent one or more traffic lanes where, for instance, there is more than one lane approaching a stop line and continuing in the same direction. A traffic lane may be represented by one or more links. In the situation where there is a single lane on the approach to a stop line containing traffic turning right, going straight on and turning left, this lane may be represented by three links. A separate link may be used to represent a particular category of vehicle, e.g. buses, even though there is not a separate lane for them. Each signal controlled junction, including Pelican crossings, is a node but a priority intersection is not a node.

The relationship between road layout and the network is illustrated in Figure 13.5.

Optimisation is undertaken by means of defining a Performance Index (PI) for each potential plan. A statistical hill-climbing technique is then used to determine minimum values of PI.

$$PI = \sum_{i=1}^{n} (W.w_i d_i + (K/100).k_i s_i$$

where N = number of links; W = overall cost per average pcu-hour of delay; K = overall cost per 100 pcu-stops; w_i = delay weighting on link i; d_i = delay on link i; k_i = stop weighting on link i; s_i = number of stops on link i.

It is necessary for the user to specify values for W and K and initial signal settings. The program will then calculate a value for PI based on these settings before altering the cycle time and working out a new value of PI. If PI_2 is less than PI_1, then it will be adopted as the new optimum, otherwise it will be ignored. This procedure is followed until a satisfactory situation is reached.

Before running TRANSYT, it is necessary to gather a large quantity of data relating to the initial signal settings (saturation flows, staging, inter-green times, minimum green times) and to flow characteristics and journey times for each link on the network. This information is then input to TRANSYT and the optimum value of PI established. Once the plan has been worked out it is necessary to implement it on a trial basis to ensure that it performs satisfactorily.

Implementation of TRANSYT in a location where there is no previous system of UTC installed is therefore a laborious and time-consuming operation that it likely to take several months. Once installed, however, there are other advantages which can be drawn.

Whilst TRANSYT establishes a series of staging plans which operate on a weekly timetable to optimise performance on the basis of

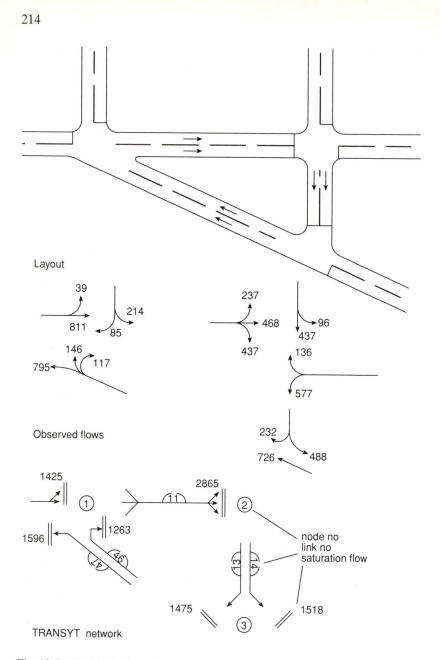

Fig. 13.5 TRANSYT: road layout and network

expected traffic flow patterns, it is likely that it will also have a central controller who can over-ride the system. Based on images received through CCTV in the control room, the controller can identify unexpected events, such as a vehicle breakdown, and institute an appropriate remedial plan. Emergency services can also be linked to the system, and a 'green wave' instituted on the path that the vehicle will follow.

Because all signals in a UTC area are linked to a central control, malfunctions are quickly identified and remedial action set in motion.

SCOOT

There is a major shortcoming with UTC operated on the basis of TRANSYT: it operates entirely on the basis of historical data. The programme optimises signal stages for the traffic conditions which occurred at some time in the past and which, on the basis of that evidence, are expected to be happening now. The controller can react to actual conditions as they occur but he does not have much more than a panic button, an ability to take emergency action if a major problem arises.

Ideally, it should be possible to short circuit the months that are involved in setting up TRANSYT so that data on traffic movements is collected automatically, fed back to the central computer, processed immediately and used to optimise the next cycle of the signals. This would be a fully traffic responsive system. It is perfectly feasible but would require enormous investment in fixed links between the signals and the computer and a very large amount of high-speed processing power.

The nearest that we have reached to a fully traffic responsive system is SCOOT[5,6] (Split Cycle and Offset Optimization Technique). In a SCOOT system information on traffic flow is collected by induction loops buried in the road surface and relayed via landlines to the central computer. The computer is thus informed of the lengths of queues building up at each signal and will utilise this information to determine the delay and stops that would result if the existing signal settings were retained, and whether any advantage is to be gained by altering the settings. If there is some advantage, the settings can be altered for the next cycle. SCOOT will thus respond to changes in traffic conditions in one cycle time, typically between 60 and 90 seconds.

The main shortcoming with SCOOT is in the automatic collection of traffic data. Induction loops can identify whether traffic is stationary, moving or absent, so they are able to gather information on queue formation. The system is still dependent on assumed values of journey speed and delay. It does, however, seem to have operational advantages over TRANSYT, and most new systems being installed are SCOOT.

References and further reading — Chapter 13

1. Institute of Highways and Transportation. *Roads and traffic in Urban Areas: Urban Traffic Control.* 160–169. HMSO. London. 1987.
2. Vincent R A, Mitchell A I and Robertson D I. *User Guide to TRANSYT version 8.* TRRL Report LR888. Transport and Road Research Laboratory. Crowthorne. 1980.
3. Transport and Road Research Laboratory. *TRANSYT – A traffic network study tool.* TRRL Report LR 253. Crowthorne. 1969.
4. Department of Transport. TRANSYT version 9, *Users' manual. TRRL Application Guide 8.* Transport and Road Research Laboratory. 1988.
5. Department of Transport. *SCOOT – the UK traffic responsive method of co-ordinating signals.* TRRL leaflet LF1025. Transport and Road Research Laboratory. Crowthorne. 1981.
6. Hunt P B, Roberston D I, Bretherton R D and Winton R I. *SCOOT – a traffic responsive method of co-ordinating signals.* TRRL Treport LR 1014. Transport and Road Research Laboratory. Crowthorne. 1981.

Chapter 14

Computer Aided Highway Design

The CAD System

One of the earliest applications of computer aided design (CAD) in civil engineering was in the layout of highways. This is because the determination of a highway alignment involves a large number of repetitive calculations in order to fix the geometry of the road and determine sufficient points to enable it to be set out on the ground.

Minor alterations to properties such as radius of curvature on a horizontal arc or length of a vertical curve may be necessary in order to fit the alignment to whatever constraints are present, and each such alteration involves completely reworking the geometrical calculations. It may also be that the calculation of earthworks volumes, a tedious procedure of simple arithmetic, are crucial to the economic viability of the scheme. Thus, highway design was seen as a suitable task for computerisation from the late 1960s.

It was not, however, until comparatively recent developments involving mini computers, PCs and high resolution graphics monitors that CAD became feasible even for small projects and the full power of the computer to create on-screen visual images began to come into its own.

Highways applications for computer systems can be conveniently divided into two groups. Major highway projects are nearly always designed using IMOSS, while estate layout can be conveniently carried out using one of a number of computerised draughting packages which produce layout drawings in much the same way as a draughtsman might do on a drawing board. Railway alignments have many similarities with highways and can be computer generated using adaptions of the same programs.

When using a CAD system, it is important to remember what it is that the computer is doing and, equally important, what it is not.

All a computer can do is to carry out vast numbers of simple calculations very, very quickly. It cannot be creative, nor can it pass judgements (unless it has already been programmed to react to a pre-defined situation in a pre-defined way, which might look like a judgement). It can only do what it has been programmed to do, and the responsibility for the output must always belong to the operator.

A CAD system will normally consist of a number of components:

1. Input device(s), which might include keyboard, tablet, scanner, light pen and mouse. These are the means by which the operator communicates instructions and/or data to the computer.
2. Processor. This is the 'black box' (or more commonly a white one) which contains the microelectronics by which the computer performs its logic.
3. Output device(s): primarily a VDU screen through which the results of the computer's calculations are communicated to the operator. For CAD applications, the screen will be a high resolution graphics screen capable of transmitting high quality visual images. Output can also be downloaded to a printer or a plotter, usually as a separate operation, so providing hard copy of the computer's work.
4. Data storage, which may be contained on a hard disk within the computer, on floppy disks which can be physically removed and stored in a disk file or in a separate disk storage facility and file server.

In addition, a single computer might be linked to other computers in the same office, the same building or, via a modem and the telecommunications network, to any other computer similarly linked anywhere in the world. The advantage of such a network is to allow data and programs to be transmitted between different locations without the physical movement of vast quantities of paper.

Depending on whether it is primarily geared to major highway projects, or to highways design as a subsidiary operation to other activities, a typical CAD installation in a highways office is likely to consist of one of two types of installation:

1. A minicomputer workstation, either stand alone or networked, running interactive MOSS (currently version 9.0) which will be capable of producing all highway alignment calculations and drawings to UK and some overseas standards, calculating earthworks quantities and producing sophisticated three dimensional images

for use with photomontage and other techniques for public consultation. Nearly all major highway schemes in the UK are produced by local authorities or consultants using such a system. MOSS can also be linked to AutoCAD for production of drawings and to drainage design packages using the Modified Rational method of storm sewer design (MicroDRAIN, WALRUSS).

2. A network of PC compatible machines running a range of software, probably linked to AutoCAD to generate layout drawings. A wide range of specialist software is available for undertaking calculations and producing layout drawings for, for instance, junction layouts or traffic signs. A separate storage facility may be linked to the network if access to large quantities of data is required, as in the case of an assignment model or GIS database.

The economics of computerisation are continuously changing. Installing a workstation and the necessary peripherals to run MOSS would be unlikely to cost less than around £40,000. Once installed, however, there will be sufficient power in the machine to cope with almost any conceivable highway design project.

A PC network would be much cheaper and would be more directly applicable to other purposes. It would, however, be slower and a large project would have to be broken up into smaller parts, with consequent time penalties for the engineers working the system.

MOSS

MOSS is a three dimensional surface modelling system used for the survey, mapping and design of a wide range of civil engineering applications. It was originally developed by a consortium of local authorities (Durham, Northamptonshire and West Sussex) whose staff perceived the need for a new approach to computer aided highway design that did not depend simply on the computerisation of traditional methods. These are essentially restricted to the cross section design of single alignments and cannot be easily extended to allow the automated design of interchanges and urban projects. MOSS aims to design all highway projects of whatever complexity and provides an integrated, computerised solution from site survey, through design and quantities to construction drawings and setting out. (The Department of Transport developed a system called BIPS which is the traditional procedure transferred to a mainframe computer. BIPS came into being before MOSS but it is only rarely used today).

The system was subsequently purchased by some of the engineers who had been concerned with its development and it is now developed and marketed by Moss Systems Ltd, based at Horsham, Sussex. MOSS is the market leader in the United Kingdom for computerised design of major highway projects. Licensed users include almost every local highway authority, together with almost all the large consultancies specialising in highway works. The program suite IMOSS, which is the name by which the software is normally known, refers to the interactive version, whereby instructions can be given to computer by the user reacting to on-screen menus. MOSS as a whole provides interactive, linemode and batch operation, which provides exceptional flexibility for operations from on-line editing of data and interactive design to off-line production of cross-sections and final drawings.

A feature known as UPM (User Programming Module) allows users to develop their own customised menus and design procedures to include local practice and to simplify the use of the system.

Apart from the highway design work for which it was originally developed, MOSS is extensively used in the design of railway, airport and canal alignments and in pavement resurfacing and reconstruction. A recent innovation was its use in determining the line of a new metro tunnel in Singapore.

The system works by the creation and manipulation of models of surfaces, which may include the existing ground levels and any future ground levels. It therefore has applications in any situation in which ground levels will be changed as a result of the implementation of the design being prepared. The 'surface' being modelled does not actually have to be a surface in the sense that it should be bounded on one side by the open air. It can equally well be the interface between two layers of geological strata or between two layers of a pavement structure.

The preparation and manipulation of the model within the computer is sufficiently fast for it to be practical to consider a large number of alternatives for the finished design. It is possible to prepare intermediate models as an aid to the management of earth moving operations. MOSS offers flexibility and accuracy for determining earthwork quantities for linear highway projects and for landscaping and quarrying applications. Quantities may be determined by cross section techniques or by using triangular surface models created automatically from string surface models. This isopach technique produces very accurate cut and fill volumes and is ideally suited to determination of complex quantities such as those involved in road resurfacing.

This gives a further range of types of civil engineering scheme for which the use of MOSS can be advantageous, for example:

opencast mining and quarrying
landfill sites
landscaping
stockpile control
sports grounds
earthfill dams
marinas

Three dimensional surface models within MOSS are defined by strings. A string is defined as a series of points joined by imaginary straight lines (or curve fitted lines) to the point before it and the point after it. Note that the order of points is important and that strings have a direction associated with them. In processing data, you may wish to pass forwards or backwards along the string (see Figure 14.1). The direction of the string will be important when deciding which is the left hand side and which is the right hand side. This, in turn, will often be important in determining the sign to be used in the input data file.

The starting point for most MOSS jobs will be the creation of a model or models representing the existing ground and any features (trees, buildings, streams, roads, etc.) which are relevant to the design. A wide range of options exists within MOSS for inputting the initial

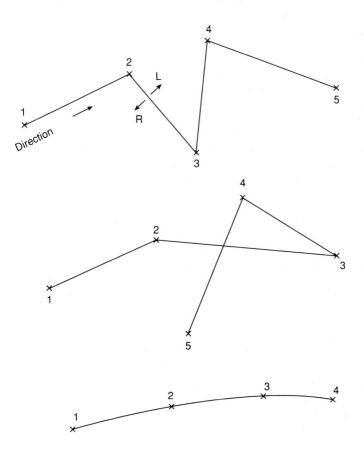

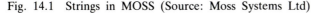

Fig. 14.1   Strings in MOSS (Source: Moss Systems Ltd)

model, including processing raw survey data gathered in the field, inputting fixed points of known co-ordinates or inputting digitised data supplied by, for instance, Ordnance Survey. The MOSS site measurement module provides for the automatic recording of data by portable computer. Data can then be directly downloaded into MOSS to produce the surface model and survey drawing.

All strings within MOSS are labelled using four alpha-numeric characters, the initial character identifying the type of string or feature. Thus, M001 is a Master Alignment String and PSSA is the Stations' String, linking the stations of a traverse. Subsequent master alignment strings might be called M002, M003 etc., but PSSA is a standard label which always refers to the string defining survey points on a traverse. The initial letter 'P' actually refers to a 'point', string, i.e. a string defining a series of distinct points which are not joined. PLP1 might be a string defining the position of lamp-posts. If you want to label the north channel, you might well call it CN01, and the south channel CS01. Other examples of the use of a standard initial letter are 'B' for building, 'F' for Fence, 'D' for Ditch. This convention is particularly useful for surveyors logging data in the field, but it is also important that designers follow it so as to avoid confusion.

Within MOSS there are a large number of MAJOR OPTIONS and MINOR OPTIONS which are operations to be performed on a model or models. The user selects the major option(s), model(s) and minor option(s) with which he wishes to work and inputs data through a coding system defined for each major and minor option. In the interactive version of the program, this coding system is reproduced through the hierarchy of on-screen menus. It is not, strictly speaking, necessary for the user to be familiar with the major and minor options that he is using, but such familiarity makes the procedures easier to follow.

The overall structure of data input is:

MAJOR, MODEL, MODEL
MINOR, data
MINOR, data
999

where MAJOR is a major option name, MINOR is a minor option name, MODEL is a model name, 999 is a standard code to indicate the end of a set of minor options within a major option.

MAJOR OPTIONS have names consisting of letters which usually relate to the function which they undertake. They are all defined in the program and listed in the manual.[1,2] For instance, SURVEY, HALGN, and VCUSP are all major options (for inputting and processing survey data, fixing a horizontal alignment by means of circular arcs and transitions and fixing a vertical alignment by means of a cubic spline respectively).

MODEL NAMES can consist of up to 28 alphanumeric characters.

It is recommended that the model name should relate to the model it identifies. The system can work on more than one model at a time. It may, for instance, be desirable to store the existing ground surface in one model, and the proposed ground surface in another. The calculation of earthworks volumes will require work on both models.

MINOR OPTIONS are defined by means of three numbers, some of which are specific to the major option concerned and some of which are general and can be used at various points through the program. These Global Minor Options include 017 and 018 which can be used to change the default system parameters, 019 which defines selection masks and 900 which calls up a macro.

It may be necessary, or at least convenient, to input data relating to some major options, in LINEMODE. This involves data input through the keyboard. Data is arranged in a series of fields. Each minor option line has ten fields and data must be typed in to the right field. In the manual, all the fields are listed for each minor option. Each field has to be separated by commas, although where a large number of fields are blank, you can avoid a long row of commas with nothing between them by using the form N=, where N is the field number to be identified. Off-line (BATCH) processing is available and may be the most convenient way of producing certain output, such as a large number of sets of cross sections.

A file of raw input data might look a bit intimidating. It relates to the time that it was necessary to use card readers to input MOSS to a mainframe computer. The sheet is set out to follow the convention of the card reader. The data input system used when typing files directly through a workstation follows the same convention.

A file of commands to put into MOSS might look something like:

DESIGN, EXAMPLE CURVE
THIS WILL PUT A NEW STRING M005 JOINING TWO EXISTING
STRINGS, CNO7 AND CWO4 INTO THE MODEL EXAMPLE CURVE
145,CN07,CWO4,M005,-10,1,0,-9.99,,,9.99
999

In this example:

DESIGN is a major option
EXAMPLE CURVE is a model name
The three lines starting THIS WILL PUT. . . is just a note inserted into the data to remind the user what he is doing. This can always be done by putting in a minor option code 000 or by simply leaving the space blank.
145 is the minor option – it actually generates a circular master string between two previously defined strings
CN07, CW04 and M005 in Fields 1, 2 and 3 are string labels. Fields

1 and 2 define the first and second strings to be joined. Field 3 is the new string to be generated – it is a master string, so its label must begin with 'M'. This convention runs through MOSS.

Fields 1 and 2 usually contain string labels that are to be used as reference data. Field 3 usually contains the label of a string that is to be created or amended.

Field 4 in this minor option defines the radius of the string to be created. The negative sign indicates a left hand curve when travelling forwards along the string from CN07 towards CW04

Field 5 indicates the chainage interval, i.e. the distance along the string being created between chainage points that will be created. Field 6 indicates that chainage of the start of the curve, i.e. the point at which the new string is tangential to the string shown in Field 1.

Fields 7 and 10 are the offsets of the centre of the circular arc which will be defined by the new string from the first and second strings respectively. They are usually marginally (0.01) less than the radius so as to ensure an intersection between the existing and new strings. This is needed for future editing. They will normally have the same absolute value but may be of different sign. If Field 10 is left blank it will be assumed to be the same as Field 7.

Fields 8 and 9 are empty.

We could have equally well put in this line as:

145,CN07,CW04,M005,-10,1,0,-9.99,10=9.99

Identifying a field by 10= can save long lists of commas, which are easy to get wrong

999 in the example shows that this is the end of a major option set

In order to simplify the process of typing in data, some major options have sets of field descriptors associated with them. So you might want to put in something like FC=2500 to indicate 'Finish Chainage 2500' instead of worrying about which field that particular piece of information has to go in. Complete lists of field descriptors are given in the manual under the major option headings to which they refer. Major option DESIGN does not have field descriptors associated with it, although a lot of the processes involved in creating and amending new strings relative to existing ones can be done interactively.

Systems to run on PCs

The following section provides some brief information on PC-based software for highway design. Addresses for the suppliers listed are contained in the 'References and further reading' section below.

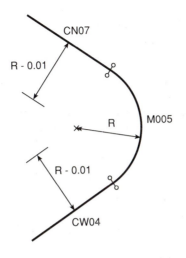

Fig. 14.2   MOSS: Major option DESIGN; minor option 145 (Source: Moss Systems Ltd)

Eclipse

Eclipse, which is developed and supplied by Ground Modelling Systems Ltd[3] is an interactive ground modelling package which consists of series of modules concerned with surveying, mapping, civil engineering design and drafting.

Eclipse operates on industry standard IBM-PCs and compatibles. Input of survey data can be by direct entry through the keyboard, from the Eclipse Field System using a Husky Surveyor data logger, by digitising existing drawings or by digitised data from Ordnance Survey.

Design is undertaken by referring proposed points to existing survey detail, so setting out data is generated by reference to existing survey control stations. Horizontal and vertical alignments are generated by defining intersection points or by sketching elements freehand. Estate layouts can therefore be generated rapidly in a manner quite similar to that which would be used by an engineer working at a drawing board.

More complex three dimensional models using 3D strings can be built from the database of points and string features. This may be used to produce contours, isometric and perspective views and volume quantities.

Eclipse can be linked to other packages including AutoCAD and MOSS GENIO. Drainage design can be undertaken interactively and plan, long sections and manhole schedules produced. Full sets of contract drawings can be produced

Personal design systems

Personal Highway Designer (PHD2) is one of a range of systems produced by Elstree Computing Ltd,[4] a member of the Laing Group. It requires the use ECL's integrated Personal Ground Modeller System (PGM2). It will run on industry standard supermicrocomputers.

Design starts from a display of the site generated from PGM2. The user can store fixed parameters relating to design speed, transition curves, permissible gradients, superelevation, critical radii and carriageway widths.

The command INSERT STRAIGHT inserts a straight element which may be fixed (passing through two defined points or passing through one defined point on a known bearing), floating (passing through one defined point or having a known bearing) or free. A straight may be tangential to one or two circular curves. Transitions can be fitted within the program.

The command INSERT L(R) CURVE inserts a left (right) hand circular curve which may be fixed, floating or free and may be tangential to one or two other elements. Transitions can be fitted within the program.

Vertical alignments can be generated on an established horizontal alignment using the commands GENERATE PROFILE, INSERT GRADE and INSERT VCURVE. Vertical curves are parabolic.

Horizontal and vertical alignments are automatically combined to form a three dimensional model from which cross sections can be plotted and earthworks quantities calculated. Junctions can be designed by the interactive generation of alignments, features and surfaces.

Drawings and quantities can be generated from PHD2 which can also interface with AutoCAD. A new version of the program, which will run inside AutoCAD, is understood to be due for release shortly.

Highway design from GTS

M3 is the highway design suite developed by GTS Ltd,[5] who market a wide range of civil engineering software covering structural analysis, steel, reinforced concrete and masonry design, drainage design using the modified rational method (MicroDRAINAGE) and ground engineering.

The system can use a digital ground model created either at the keyboard or captured on site using a suitable datalogger. The horizontal alignment module uses conventional tangent, clothoid and circular arc to produce centre line information at user defined chainage points. The vertical alignment module produces a series of straight grades and parabolic curves to define levels on the centre line. A link to a cross section database and ground model enables the production of cross sections and earthwork quantities. Output can be in the form of tabulations or, if linked to AutoCAD, drawings.

Alignments can also be produced through the ProSURVEYOR package, also marketed by GTS. ProALIGNMENT provides facilities for designing linear structures such as roads, pipelines and channels. Horizontal and vertical alignments are defined using either freehand interactive element insertion or co-ordinate definition, giving start and end conditions of the alignment elements. A section template technique is employed to provide lateral dimensioning.

References and further reading — Chapter 14

1. Moss Systems Ltd. *User Manual for MOSS V.9.0*. Horsham. 1991.

New developments in CAD inevitably happen quickly. Software suppliers are therefore keen to respond to requests for information on the latest versions of their products, and the following details might be helpful:

2. MOSS Systems Ltd., Moss House, North Heath Lane, Horsham, West Sussex. RH12 4QE. Tel: 0403 59511. Fax: 0403 211493.
3. Ground Modelling Systems Ltd., Rockingham Drive, Linford Wood, Milton Keynes MK14 6PD. Tel: 0908 667799. Fax: 0908 665171.
4. Elstree Computing Ltd. Elsmere House, 12 Elstree Way, Borehamwood, Herts. WD6 1NF. Tel: 081-207-2000. Fax: 081-207-5910; and Cornbrook House, 1 Brindley Road, Old Trafford, Manchester. M16 9HR. Tel: 061-876 4476. Fax: 061-873-7246.
5. General and Technical Systems Ltd, Brooklyn House, Brook Street, Shepshed, Leics. Tel: 0509-508060. Fax: 0509-507564.

Chapter 15

Estate Roads

A hierarchy of urban roads

Individual urban streets can be located in a hierarchy according to the function that they perform. A variety of titles can be adopted, but the following are amongst the more common:

Primary distributor
District distributor
Local distributor
Access road
Access ways
Mews court
Private drives
Cycleways
Footways

The standards to which the road should be constructed follows from its position in this hierarchy.

General guidance on layout of residential estate roads is given in Design Bulletin 32 (DB32)[1] originally published by the Departments of the Environment and Transport in 1977 and republished in a revised edition in 1992.

Estate roads, except those defined as private drives, will normally be adopted by the local highway authority and thus become maintainable at the public expense. The authority is therefore concerned to ensure that the construction standards adopted are appropriate and will not be the cause of excessive expenditure in the future. All highway authorities publish their own guidelines to developers to act as supplements to DB32. It is for the authority to determine the extent of detail included in these guides, and the frequency with which they are updated. For instance, Kirklees Metropolitan Council uses the former West Yorkshire County Council Guide,[2] which is very prescriptive as regards both layout and specification. The developer has little alternative but to look up the appropriate part of the Guide and adopt the required solution. Arguably, this approach discriminates against innovation and produces unnecessarily expensive outcomes.

A less prescriptive guide, more recently updated, has been produced by Surrey County Council.[3] This pays particular attention to the need for speed restraint on residential roads and provision of parking.

There have been some changes in public perception of the role of estate roads in parallel with the need to adopt traffic calming measures on existing streets. The 1992 edition of DB32 has taken note of this, with additional emphasis being given to the needs of cyclists and pedestrians.

Industrial estate roads[4] can also be developed on a hierarchical principle although they are unlikely to be lower down the hierarchy than Access Road. Pavement structures will need to be stronger and turning facilities less restricted than on a residential estate. Roads giving access to a single industrial or commercial premises, or to a modern light industrial estate or office park, are unlikely to be adopted by the highway authority.

The designer of the estate road is not only concerned with operational functions. He is also concerned to enhance the amenity of the area, and the concept of minimising danger and restricting traffic speeds was inherent in the original version of DB32, long before anybody in this country had adopted the idea of traffic calming.

Distributor roads

In a purely theoretical urban network, of the form shown in Figure 15.1, all roads could be identified as being 'distributors', 'access roads' or 'residential'.

In such a system, all trips would start from premises located on residential/access roads, proceed up through the hierarchy of distributor roads for the main part of the journey and then go back onto access/residential roads to reach their destination. Such a system cannot exist in practice, although something approaching it can be

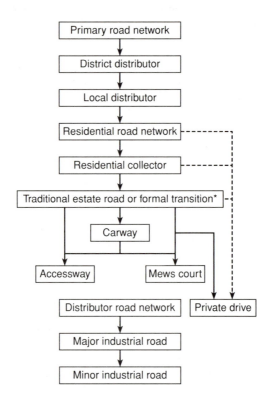

Fig. 15.1   The urban road hierarchy

found at Milton Keynes where there is a clearly defined system of distributor roads and where residential, commercial and industrial areas are easy to identify. However, it is possible to define the function of any road by reference to the position in such a hierarchy that it comes nearest to occupying and hence to determine the standards to which it should be constructed.

Distributor roads are the highest level in this hierarchy. The category may be sub-divided, primary distributors being the main cross urban links, district distributors linking one identifiable area with another and local distributors taking traffic across a single area.

The main function of a distributor road will be to carry through traffic, and it should therefore have very little direct access from adjacent properties. This may be the most difficult aspect to reconcile when defining an existing road as a primary distributor, although the provision of service roads may be possible.

The road will be designed by a similar process to that used for highway link design on rural roads. As an urban road it is likely to be subject to a speed limit of either 30 mph or 40 mph, which might be considered to translate into design speeds of 48 km/h (13.3 m/sec) or 64 km/h (17.8 m/sec) for the purpose of determining sight distances and alignment details.

Road width will normally be single or dual carriageways 7.3 m wide. In an ideal situation, pedestrians and cyclists would be carried on a separate network so there would be no footway immediately adjacent to the carriageway except where required to give access to bus lay-bys.

Access roads

Within the access road band, there will again be a hierarchy of separate road types, with names determined by the particular local authority design guide. West Yorkshire used the names Residential Collector and Traditional Estate Road. Surrey uses a system of 'Design Values' (DV) which is developed from the number and capacity of the buildings served and is therefore an approximation of comparative traffic flows. For our purposes, the term 'Access Road' includes a road which may well have residential property fronting directly to it but will also serve minor residential roads and cul-de-sacs. The total number of properties served by it will be in the range 15–300. The description includes industrial estate roads serving light industrial estates, office parks and retail developments.

It is more appropriate to consider a target maximum speed (TMS) than a design speed in this situation, as restriction on speed is a specific objective. At the upper end of the range a TMS of 50 km/h is appropriate, whereas at the lower end 30 km/h should be used. Carriageway widths will be between 4.1 m and 6.0 m. Various techniques can be used to ensure that the TMS is not exceeded, including careful landscaping, use of parking areas to ensure that the path of through vehicles is tortuous, inclusion of sharp bends and speed restraint points where carriageway width is narrowed to 3.0 m. It is important that any measures to restrain speed should be dicussed in advance with the emergency services.

At the lower end of this range, it may be appropriate to use shared surfaces as a means of reducing speed and enhancing the visual amenity. In general, however, these roads will have upstanding kerbs and adjacent footways at least 1.8 m wide.

Particular care needs to be given to parking arrangements. Where garages are included within the curtilage of the building, the requirement is usually for one additional parking space for every garage space. There is still a need for unassigned parking places to be provided within the public highway for visitors and delivery vans.

Local highway authorities will lay down standards required based on their local knowledge of the area, but a typical standard is shown in Table 15.1

Where parking spaces are not assigned to a particular dwelling, it is important to ensure that they are conveniently located, otherwise they will not be used. Where walking distances are excessive, the result tends to be that vehicles are parked on verges leading to damage to kerbs, manhole covers and the verge itself.

Perceptions of the standards required for parking are changing. Traditionally, the objective was to get as much parking as possible off the public highway in the interests of keeping traffic moving. Today, however, parked vehicles, properly located, are seen as a potential means of keeping down speeds in the interests of safety.

Table 15.1 Typical parking standards

| | Number of parking spaces per dwelling | | |
Type of dwelling	2+ bedroom	1 bedroom	OAP
1. No assigned parking	1.5	1.0	0.5
2. Some assigned parking	1.75	1.0	0.6
3. Grouped garages	2.25	1.75	
4. Single garage	2.5		
5. Double garage	4.25		

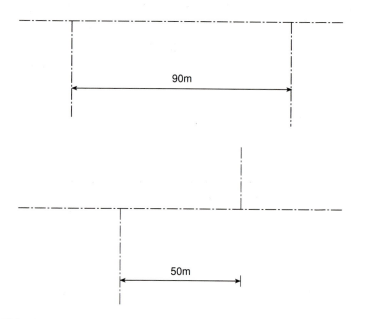

Fig. 15.2 Staggered junctions on an estate layout

Junctions on access roads should be carefully located with a minimum distance of between 45 m and 90 m between the centre lines of side roads on the same side of the main road, and between 25 m and 45 m between staggers. (See Figure 15.2). Kerb radius at junctions between access roads and distributor roads should generally be 10 m unless a significant number of long vehicles are expected to use the junction. Kerb radii should be restricted on purely residential roads as a means of reducing speeds, although a judgement will have to be made bearing in mind the damage caused by vehicle wheels striking the kerb.

Sight lines at junctions are fixed by means of the X– and Y– distances (See Figure 15.3) and depend on the design speed/TMS on the major road. Suitable standards are shown in Table 15.2.

Residential roads

Many of the considerations which apply to access roads will also apply

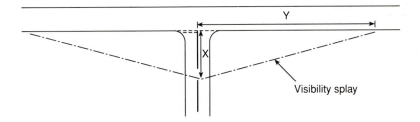

Fig. 15.3   Priority junctions: X- and Y-distances

Table 15.2   X- and Y- distances for visibility splays at junctions

Major Road Design Speed/TTS (km/h)	X (m)	Y (m)
85	9	160
70	9	120
60	9	90
50	5	90
40	5	40
30	2.5	40
20	2.5	20

to residential roads. These can generally be regarded as serving between 1 and 15 dwellings, although at the lower end they will not be adopted by the local highway authority and simply become private drives. A TMS of not more than 20 km/h should apply and this will be achieved by limiting the length of a cul-de-sac to no more than 200 m. Residential roads will not normally provide through routes except in the case of a loop having both ends on the same access road.

Carriageway widths will be between 2.8 m and 4.5 m. Shared surfaces without kerbs and separate footways will be likely, although in such a situation there must be a verge or other facilities provided within the highway in order to accommodate services. Adequate facilities must be provided to ensure that any vehicle likely to be using the road can turn in order to exit in forward gear. The worst case is usually considered to be a large removals vehicle or pantechnicon, rigid not articulated and 9.5 m in length.

Pedestrian and cycleways

Early attempts to design environmentally friendly estates frequently concentrated on the desirability of separating vehicular traffic from pedestrian and cycle traffic. The Radburn layout, developed in the United States, concentrated on segregated vehicular access giving direct access to parking areas, with an entirely different network for pedestrians and cyclists. Typically, houses would be arranged in squares facing inwards on to green areas, while vehicles would gain access from the rear, on the outside of the square.

Such developments, however, were often seen as being primarily designed for the benefit of the vehicles. Pedestrian routes tended to be tortuous and to have problems with lighting, security and vandalism. At points where the pathway had to cross a road, it tended to be the path that disappeared underground into a dingy and evil-smelling underpass. Cycle routes, such as those built in the early 1960s around Stevenage, Hertfordshire, were lightly used and represented a poor return on substantial capital investment.

An attempt to overcome these problems was made with the layout design of Milton Keynes. The area of the city is divided up by a grid formed by distributor roads. Each square within the grid is relatively self-sufficient, so that journeys to local facilities can be undertaken by walking or cycling. Longer journeys, to work or to the city centre, would have to be made by car but these would be undertaken using the distributor roads. Public transport primarily follows the distributors and can be accessed by footpath from within the residential zones. A network of 'red routes' (not related to the traffic management measures being adopted in London) allows pedestrian and cycle paths throughout the city. Where these paths cross the distributor roads they do so on bridges, not underpasses, and the levels are such that the path approaches the bridge at deck level. The Milton Keynes solution, however, while generally recognised as being successful, is not available in traditional locations, where there is neither the space nor the finance available.

The advocates of traffic calming suggest that the solution lies in making traffic behave in a manner that is friendly to pedestrians and cyclists, rather than the other way round. This has implications for the design of new estates. If Milton Keynes had been designed today, the network of pedestrian and cycle routes might very well have been rather similar although the distances involved would have been shorter. A very large amount of land space in Milton Keynes is taken up by the distributor roads and the junctions between them, so the distance between one residential cell and another is substantial. Car parking, particularly in the central area, takes up an enormous area, so the cyclist, aiming to do his shopping, may have a cycle track that will take him to his destination but he will spend a lot of his time crossing distributor roads and car parks.

We clearly have some way to go before we are able to claim that we have designed a city, or even an estate, around the need to be friendly to pedestrians, cyclists and users of public transport. The needs of the motor car still take priority in the majority of planning decisions. We are, however, beginning to realise that an alternative approach is possible. How quickly this change takes hold, or whether it takes a hold at all, will depend on public reaction to schemes now being implemented and its reflection in the political, decision-making process.

References and further reading — Chapter 15

1. Department of the Environment/Department of Transport. *Residential roads and footpaths: layout considerations.* Design Bulletin 32. HMSO. London. 1992.
2. West Yorkshire Metropolitan County Council. *Highway Design Guide Parts 1 to 5.* Wakefield. 1985.
3. Surrey County Council. *Roads and footpaths: a design guide for Surrey.* Guildford. 1989.
4. English Estates *et al. Industrial and commercial estates: planning and site development.* Thomas Telford. London. 1986.

Chapter 16

Safety Engineering

Road safety audits

It has long been standard procedure in local authority highways offices to maintain records, drawn up in association with the police, of accidents which occur on roads for which they are responsible. Some details of the procedures involved are shown in Chapter 6. Data so gathered can be used to forecast levels of reduction in accidents which might result from road improvements, and this reduction is an important component of the economic appraisal of individual schemes.

However, this approach, while it may be essential, is by no means sufficient. A common cry from members of the public seeking to influence highway authorities is 'do we have to wait until someone gets killed before you do something about this dangerous piece of road?' Too often, the answer is 'yes', because we do not know enough about the circumstances which contribute to the occurrence and the seriousness of road accidents. The standard classification by level of injury may have more to do with the age and state of health of the occupants of a car involved than it has with the design of the road where the accident occurred. On the other hand, misleading information provided to the driver, which may affect the speed at which he is travelling, could equally well be a major contributor to the level of injuries sustained.

Too often, the investigation of accidents is concerned to find fault in the behaviour of one or all the drivers involved. We have, in the

United Kingdom, an extraordinarily complex system of civil courts where very powerful insurance companies can battle for years in the hope of proving that it was the other man's fault. In fact, of course, road accidents are rarely the fault of one person. A large number of contributory factors are likely to have come together, a change in any one of which would have prevented the accident. One such factor is likely to be the layout of the road. This is not to say that there was anything wrong with the design or that the accident was, in part, the fault of the designer. It is simply to say that, had circumstances been different, the accident might not have happened. The role of the designer is not to ensure that, if anything goes wrong, he can find somebody else to blame. It is to ensure that he does everything in his power to ensure that nothing goes wrong.

The Institution of Highways and Transportation can claim much of the credit for the development of this philosophy. In 1986, they published guidelines on accident prevention.[1] The cause was taken up by the Government who included a requirement in the Road Traffic Act 1988[2] to the effect that local authorities, in constructing new roads, must '. . . take measures such as appear to the authority to be appropriate to reduce the possibility of accidents when the roads come into use'. The Government also declared their intention to reduce the number of casualties due to road accidents by one third by the year 2000.

A safety audit is intended to satisfy, at least in part, this requirement. It is a procedure which an authority has to go through as part of the design process. Guidelines, which include details of the objectives and principles of audit, have again been provided by the Institution of Highways and Transportation[3] and the Department of Transport have published a standard[4] and advice note.[5]

The word 'audit' is used deliberately in order to give greater emphasis than the word 'check'. The latter is familiar to civil engineers as a mechanism to ensure that calculations have been undertaken accurately. It would not normally be taken to apply to the principles behind a design.

Audit implies that a scheme is considered systematically and at various stages in its development by a safety specialist acting in concert with the designer. It includes the allocation of responsibility for implementing its findings. Check lists are contained in both the IHT Guidelines and the Department of Transport Advice Note for assisting audit teams in carrying out their function, although it will probably require some years of experience to determine whether the checklists as published are adequate. At present, they should be regarded as an aid rather than as a complete definition of the work to be done in carrying out the audit.

Management of human resources involved in audit needs to be appropriate to the scheme in question. IHT guidelines provide six alternative structures:

1. A specialist advice and audit team within the organisation undertaking the design, which has to undertake the audit and issue a safety certificate before work can proceed.
2. A specialist advice and audit team reporting to a third party, who may be a senior manager of the organisation with no direct responsibility for the design, or may be the client's project manager. In a DTp scheme, the design team would be within the main consultant commissioned for the work, the specialist team could be within the same consultancy or a different one and the report would be made to the Department's Project Manager who would authorise progress.
3. A specialist advice and audit team reporting to the designer, with the designer deciding on action and progress.
4. A second design team carrying out the audit and reporting to an independent assessor.
5. A second design team carrying out the audit and reporting to the first design team.
6. The design team carrying out its own audit and preparing sufficient documentation to meet the requirements of the audit. There are probably not many circumstances when this could be regarded as being adequate.

The IHT Guidelines suggest that audit should be carried out at four stages during scheme preparation, although on smaller schemes it may be possible to combine the second and third stages:

1. At feasibility/initial design stage, in order to influence route choice, standards to be adopted, junction provision and impact on the existing network.
2. On completion of draft plans or preliminary design.
3. On completion of detailed design but before preparation of contract documents.
4. Immediately prior to opening of the scheme.

Many potential hazards occur because of inadequate liaison between different organisations or different teams within the same organisation. Audit procedures therefore need to be established when work is being carried out on the highway by developers or by statutory companies.

It may be appropriate to establish audit procedures on existing roads, thereby identifying projects for future implementation. Many local authorities are now establishing Safety Engineering Units whose role will not only be to monitor accident rates and carry out audits within the authority but to identify, and possibly design, potential schemes whose major purpose is to reduce accidents. The establishment of such a unit is likely to significantly influence the policies and priorities of the authority.

Traffic calming

Traffic calming refers to the adaption of existing road layouts in order to reduce the speed of vehicles travelling through areas where they are likely to come into conflict with other road users. It operates through application of techniques such as road humps, road narrowing, chicanes, management of on street parking, traffic gates and, possibly, statutory measures such as speed limits and priority definition. These techniques are described in detail in Chapter 24.

One of the primary objectives of traffic calming schemes is the reduction of injury accidents. The potential is considerable since the majority of road accidents occur on urban roads. By reducing the speed of vehicles it not only reduces the likelihood of an accident occurring, because a driver has more time to take avoiding action, but it also reduces the severity of the injuries which result. A pedestrian struck by a vehicle travelling at 40 mph has almost no chance of survival, whereas 50% of pedestrians struck at 20 mph survive.

It is therefore likely that a primary duty for the safety engineering units now being set up will be to identify localities where traffic calming measures might lead to reductions in accident rates.

Safety at roadworks

The existence of work being carried out on the highway significantly increases the danger of accidents. Additional obstructions in the form of cones, barriers, temporary works, plant and excavations are necessarily in place. The condition and layout of the road will be unfamiliar to drivers. The capacity is likely to be reduced, leading to a greater likelihood of congestion, standing traffic and driver frustration. And people concerned with supervising the works and carrying them out are brought into situations of potential conflict with moving traffic.

Appropriate measures to be taken when ensuring safety at roadworks are detailed in Chapter 8 of the Traffic Signs Manual.[6] There is also a brief guide which forms a companion volume and is intended as a full-colour, simplified version for the benefit of foremen, supervisors and inspectors, although there have been problems with some of the oversimplification contained in this booklet and it has been withdrawn pending republication in late 1992 or early 1993 in the form of a code of practice required by the New Roads and Street Works Act 1991.

'Roadworks' are defined as being *any* works or temporary restrictions which cause partial or total obstruction of the highway. The highway is all the area between the highway boundaries, including verge, footway, cycleway or carriageway. Roadworks might thus be a full scale contraflow during reconstruction of a length of motorway down to opening the cover of a stop cock in an urban footway.

The responsibility for the works rests with the person or organisation carrying them out, who may be the highway authority, a statutory undertaker or body undertaking work in the highway under licence. The highway authority also has an overall responsibility to ensure that the works are properly and safely carried out. The Health and Safety at Work etc. Act 1974 (and the Health and Safety at Work (NI) Order 1978) applies and places a wide ranging responsibility on all clients, contractors, employers and employees to establish and maintain safe systems of work. A summary of the present position regarding the application of the Health and Safety at Work Act has been provided by Anderson.[7]

The objective, as defined in Chapter 8, is to achieve a satisfactory balance between getting the work done as quickly and safely as possible, optimising work efficiency and minimising congestion, delay and inconvenience. There is a dilemma here, for this implies a compromise between safety and operational efficiency which is something the Health and Safety Executive would never sanction.

The European Commission is active in the area of Health and safety at Work and has published proposals[8] for the establishment of a European Agency to:

1. set up and co-ordinate a network for the exchange of information and experience;
2. organise training courses and exchanges between member states;
3. promote co-operation in monitoring performance;
4. co-operate with other international organisations and institutions within member states.

The legal requirements contained in Chapter 8 are set out below.

1. The person (or organisation) responsible for the works must take such steps as are reasonably necessary to protect persons legally using the highway from personal injury and their property from damage. They must erect such warnings, barriers and other measures as are necessary to discharge this obligation, and must ensure their removal when the works are complete.
2. The highway authority has the ultimate responsibility for administration of all works that take place on the highway.
3. The Health and Safety at Work etc. Act 1974 applies. In particular, there is a requirement that construction personnel and plant should, at all times, be segregated from passing traffic and high visibility clothing to BS6629:1985 should be worn.

The chapter defines three 'Types of Works', which could be more appropriately described as 'Types of Signing' dependent on the manner of undertaking the works, rather than the nature of the works themselves.

- Type A works are those works which remain in operation in all traffic flow and visibility conditions. The majority of signing systems for construction operations will be Type A. Types B and C are likely to be relevant during maintenance operations such as gully emptying and mobile operations (including putting cones out for Type A works), or for works which are of short duration such as minor statutory undertakers' operations. The same scheme might involve different types of signing at different stages in the operation, which is why it is misleading to refer to types of works rather than types of signing.
- Type B works are those works which are allowed to remain in operation only when the traffic demand is less than the available carriageway capacity when the works are in place, and when visibility is good.
- Type C works are similar to Type B but most traffic signs used will be vehicle mounted.

It may be desirable to consider whether it is possible to downgrade works from Type A to Type B at the planning stage. For instance, operations such as the final reinstatement of a trench line may only take an hour or two. If it is carried out during the peak hour, it may be Type A because it involves restricting available capacity to below the expected flow. The same work three hours later may be Type B.

Works in the carriageway or close to its edge will cause restrictions to the width available to traffic, and, hence, to the available capacity. Plant operation and stored materials will also restrict width, while temporary scaffolds or access platforms may restrict height. Minimum running lanes and clearances must be provided in accordance with the requirements of the chapter, according to the type of road and traffic.

Where it is not possible to satisfy either the capacity or running width/clearance requirements because of the physical restrictions of the site, then it may be necessary to close the road for the duration of the works. Road diversions may be implemented by the highway authority by making an order under Section 14 of the Road Traffic Regulation Act 1984.

In making an order, the highway authority must pay due regard to:

1. The suitability of diversionary route for expected traffic flows.
2. Installing comprehensive direction signing at all times that the diversion is operational throughout the diversion and its approaches.
3. The need to change or emphasise priority at junctions.
4. The effect of diverted traffic on the environment.
5. Any railway level crossings or low bridges on the diversionary route.
6. Traffic signal settings.
7. Effect on loading and waiting regulations on the diversionary route.[9]

8. Effect on frontagers on both the diversionary route and the closed road.
9. Effect on public service vehicles and bus stops.

The highway authority may wish to discuss diversions with the police, public transport operators, railway operators, managers of business premises affected and other interested parties.

It may also be desirable to implement speed limits past the site of the works.

Mandatory speed limits made by order under Section 14 of the Road Traffic Regulation Act 1984 must be imposed for Type A works exceeding seven day durations on motorways or trunk roads. They should be used in other similar situations. Temporary mandatory speed limits should be at least 20 mph lower than any permanent mandatory limit.

In some circumstances, it may be advisable to introduce advisory speed limits in circumstances where mandatory ones are not justifiable. The chapter, however, points out the limitations of advisory speed limits in reducing drivers' speed below that which the driver would normally adopt when confronted with roadworks.

Lead vehicles can be used as a method of ensuring the reduction of traffic speeds past the site of the works in situations in which it is not possible to provide adequate clearances. Within the chapter, this is permitted under sections 2.5.3.4 and 2.5.2.2, whereby safety zones can be dispensed with provided traffic speed is reduced to less than 10 mph.

A description of procedures involving use of lead vehicles and careful timing of major works to resurface the A580 East Lancashire Trunk Road in the Metropolitan Borough of Knowsley, Merseyside, is given in reference 9.

Positive traffic control can be undertaken by the following means:

1. Police supervision in emergency only.
2. Manually operated STOP/GO boards on single carriageway roads in which site length and traffic conditions listed in Table 16.1 apply.
3. Signal control, using three aspect portable traffic signals incorporating vehicle actuation and the ability to operate on fixed time

Table 16.1 Conditions for the operation of manually operated STOP/GO boards

Site length (m)	Maximum 2-way flow (veh/hr)
100	1400
200	1250
300	1050
400	950
500	850

settings and manual control. Notice should be given to the highway authority before the installation of signals. The length of the site should be less than 300 m and traffic flows not so heavy that the restricted length of road becomes overloaded.

4. Priority control where ALL the following conditions are met:
    - there is clear visibility from a point distance D_1 in advance of the start of the cones to a distance D_2 beyond the last of the cones, where D_1 and D_2 depend on the applicable speed restriction as follows:

Speed restriction (mph)	D_1 (m)	D_2 (m)
30	60	60
40	70	70
50	80	80
60	100	100

- 2-way traffic flow not greater than 850 veh/hr.
- site length, measured from first cone to last cone, less than 80 m.

All-red settings are a function of site length and are shown in Table 16.2. Signals will remain green on a particular approach until the sensors indicate that no further vehicles are approaching or until the prescribed maximum green time is reached. Maximum green times can also be determined as a function of site length and are shown in Table 16.3.

Table 16.2   Determination of all-red period at automatic temporary traffic lights

Site length (m)	All-red period (secs)
<50	5
50–99	10
100–149	15
150–199	20
200–249	25
250–300	30

Table 16.3   Determination of maximum green times at automatic temporary traffic lights

Site length (m)	Max green setting (sec)
30–74	35
75–134	40
135–194	45
195–300	50

References and further reading — Chapter 16

1. Institution of Highways and Transportation. *Guidelines for accident reduction and prevention*. IHT. London. 1986.
2. Road Traffic Act 1988. HMSO. London. 1988.
3. Institution of Highways and Transportation. *Guidelines for the Safety Audit of Highways*. IHT. London. 1990.
4. Department of Transport. Departmental Standard HD19/90. *Road Safety Audits*. Department of Transport. London. 1990.
5. Department of Transport. Departmental Advice Note HA42/90. *Road Safety Audits*. Department of Transport. London. 1990.
6. Department of Transport. Traffic Signs Manual: Chapter 8. *Traffic safety measures and signs for road works and temporary situations*. London. HMSO. 1991.
7. Anderson J M. Managing safety in construction. *Proceedings of the Institution of Civil Engineers*. 92.Aug:123–32. August 1992.
8. Elder D. *Setting a safe pace – a report on resurfacing A580 East Lancs Road by Knowsley Metropolitan Borough Council as Agent Authority for the Department of Transport*. The Surveyor. 177.5190. April 1992.
9. Commission for the European Communities. *Proposal for establishing a European agency for safety and health at work*. Brussels. 1991.

PART IV

TRANSPORT AND THE ENVIRONMENT

Chapter 17

Environmental Objectives

UK Government policy

Rising political consciousness about environmental issues has been one of the features of the 1990s. In several European countries, Green parties have begun to make some sort of impact, starting in West Germany. In Britain, the European elections of 1990 saw the Greens reach a high of 5% of the popular vote – enough to give them parliamentary representation under many constitutions, but not in Britain. Their support fell again in the 1992 General Election, but they did not go away entirely. Part of the reason for their fall in support may have been the adoption of some green policies by the major parties.

If a single event can be said to be responsible for the growing awareness of the environment, it was probably the nuclear power plant disaster at Chernobyl in the Soviet Union. Western politicians were at pains to point out that the design of the Russian plant was essentially different from those in Britain and that 'Chernobyl couldn't happen here', but their cause was hardly helped by the fact that, until Chernobyl, the worst accident that had happened had been at Three Mile Island in the United States. Extensive publicity was generated about an accident at Windscale in Cumbria in the 1950s that turned out to have been more serious than had been admitted at the time. The quest for compensation from former soldiers who had witnessed the first testing of nuclear weapons and, years later, developed symptoms consistent with exposure to radiation also became headline news.

Chernobyl effectively killed the Government's plans for a major expansion of the nuclear power industry and caused them major problems with plans for privatisation of electricity generation.

Other environmental effects started to climb to the top of the political agenda, particularly the depletion of the ozone layer due to the creation of noxious gases by a number of industrial processes, including transport and the destruction of vegetation by acid rain, largely put down to emissions from coal-fired power stations.

Global warming is similarly the result of emission of the so-called greenhouse gases, carbon dioxide, methane, nitrous oxide, low level ozone and chlorofluorocarbons (CFCs). By artificially increasing the amount of these gases above that which would naturally occur, we cause a greater amount of heat energy, naturally reflected from the surface of the earth, to be retained in the atmosphere. Hence, the temperature on the surface rises, and it is anticipated that this will lead to melting of the ice caps and a rise in sea level, as well as, possibly serious, meteorological effects. Some action has already been taken to control emissions of lead and CFCs, but road transport remains a major area of concern.

The economic imbalance between rich and poor countries was highlighted by television pictures of famine in Africa and high profile media events such as Bob Geldof's Bandaid concerts. This, in turn, was linked to the rich countries consumption of limited resources, particularly oil. To some extent, Britain had been insulated from the gathering oil crisis by the energy reserves of the North Sea. For nearly twenty years, we had been a net exporter of the 'black gold'. But it has always been known that there was not an infinite supply in the North Sea or anywhere else, and conservation rather than profligacy at least began to find its way into the attitudes of governments and the people who elected them.

In Europe, the European Commission was faster than national governments to appreciate the scale of the problem and to start to take action. The water industry, in particular, found itself the target of EEC directives aimed at cleaning up rivers and inshore waters around the coast. Decades of neglect and under investment began, at last, to be reversed.

Transport is one of the major causes of the environmental crisis. Virtually all vehicles burn fossil fuels, either directly in the engine or indirectly through the use of electricity obtained from a power station.

The main categories of substances emitted by burning fossil fuels are as follows:

1. Carbon Monoxide (CO). A highly toxic gas which is formed by the incomplete combustion of fuel; 88% of UK emissions of the gas come from road vehicle exhausts. If allowed to accumulate in a closed space it will rapidly prove fatal. In the open, it oxidises to carbon dioxide. Other sources of carbon monoxide are industry

UK emissions), domestic heating (5%) and power stations

dioxide (CO_2). Formed by combustion of fossil fuels and
of carbon monoxide; 33% of UK emissions come from
itions, 23% from industry and 19% from road vehicles.
main source is domestic heating (14%). Carbon dioxide
y absorbed by vegetation, which emits oxygen, and so
m is worsened by the destruction of the rain forests and
is of plant life.

3. Sulphur dioxide (SO_2). Produced when any fossil fuel containing sulphur is burned. Coal is the principal contributor with 71% of emissions coming from power stations and only 2% from road vehicles.

4. Oxides of Nitrogen (NO_x). Formed when anything is burned, with 48% of UK emissions coming from road vehicles and 29% from power stations.

5. Smoke particles and organic compounds, which come from burning of fossil fuels and can be further acted upon by sunlight, cause problems with respiration and damage plant life; they also cause dirt on buildings.

6. Lead. Lead is toxic and can lead to brain damage in humans. Particular problems have been recorded amongst children living near highly trafficked road junctions. Emissions have been substantially reduced by the growing use of unleaded petrol.

It is clear from the figures above that a reduction in the amount of noxious substances from road vehicles would lead to a significant improvement in environmental quality. The degree of importance attached to this varies. Some would argue that it is essential if the planet is not to be destroyed, or at least rendered largely uninhabitable, in a comparatively short time. Most governments in developed countries accept that environmental issues should be accorded a high priority, but not so high as to threaten the economic prosperity which their citizens enjoy. Governments in underdeveloped countries are likely to consider that they have more important things to worry about than global warming, like feeding a higher proportion of their population or putting down the next revolution.

In global terms, the responsibility that rests on the governments of the developed world is considerable. The West is in a particularly important position, on account of the political upheavals that have taken place in the former communist world. We are the countries that consume most, and must therefore bear the major responsibility for cutting the cost of that consumption. And we are the countries with the technological know-how and the political stability to act. If Western Europe, North America and Australasia do not advance the green cause, nobody else is going to.

So it is the policy of the British Government, and most other governments in the Western democracies, to reduce the level of vehicle emissions. This can be done by:

- altering the vehicle to make it more environmentally friendly;
- altering the use that is made of vehicles by changing travel patterns.

Approaches to achieving Government objectives

The Government has various means at its disposal in order to bring about these changes, but basically these come down to:

- exhortation
- legislation
- taxation
- a combination of any or all of the above.

In Britain, the Government has made quite a lot of progress in the first objective above, making vehicles more environmentally friendly. It has made practically none at all on the second objective of trying to change travel patterns, and many would say that the policies it has adopted have in fact made matters worse.

There have been two areas in which the government has sought to reduce the level of harmful emissions: the introduction of unleaded petrol and catalytic converters and the removal of tax advantages to users of company cars, particularly those with large engines, which use most fuel.

Unleaded petrol first became available in 1985 and from 1987 there was a price advantage of about 5% in using unleaded due to lower tax rates. From 1990 all new cars have to be capable of running on unleaded fuel.

From 1988 onwards there has been a progressive increase in the taxation levelled on both companies and employees for the use of a company car. Whilst a proportion of company car users genuinely require the car in order to carry out their job, there is also a proportion whose car is merely a perk. The larger, and therefore, the less efficient, the car, the bigger the perk. A significant proportion of people were deemed to be driving around in cars with 2 litre engines when, if left to pay for it themselves, they would make do with 1 litre or do without altogether. Tax raised from employees depends on the size of the engine, the amount of miles done 'on business' in a year and whether the company provides petrol for private mileage (including the journey to and from work). The banding system used by the Inland Revenue is very coarse, and so it all works out on a very arbitrary basis, helping some and hindering others, but, in general it can be said to have made some contribution to saving fuel and encouraging cars with smaller engine sizes. Changes in the tax rules to remove some of the anomalies are imminent, but it is difficult to see how any banding system can avoid encouraging additional mileage if this has the effect of crossing the next tax threshold.

Other measures that the government can be said to have taken or intend to take include:

- Implementation of smoke emission standards for aircraft (1986).
- Compulsory fitting of speed limiters to coaches (1989).
- Extended to new heavy goods vehicles (1992).
- Retention of maximum speed limits of 70mph on dual carriageways and 60 mph on single carriageways (1990).
- MOT test to include vehicle emissions (1991).
- Catalytic converters to be fitted to all new cars (From 1993).
- Phasing in EEC standards for sulphur content of diesel fuel (1995–97).
- Stabilisation of CO_2 emissions at 1990 levels (UK government target for 2005).

Contributions by other bodies

Manufacturers have also made a contribution to reducing fuel consumption through the manufacture of more efficient engines and vehicle bodies. Government regulations have required the publication of fuel consumption figures, based on standard testing procedures, and these have featured in marketing campaigns. In 1978, the average tested miles per gallon for new cars was about 31 miles per gallon (9.1 l/100km). By 1990, it had gone up to about 37.5 (7.5 l/100km).[1] To some extent, this reflected the increased market share of cars powered by smaller engines, but this in itself reflected manufacturing progress: if one can get the same performance out of a 1400cc engine that you used to get out of an 1800cc engine, then your fuel consumption goes down. However, it is still a matter of frustration to those promoting green policies that the motor trade still emphasises speed and acceleration as selling points, rather than fuel economy.

Catalytic converters combine the carbon monoxide and nitrates in exhaust fumes to produce nitrogen and oxygen and then further oxydise the carbon monoxide and hydrocarbons to produce carbon dioxide and water vapour. Catalytic converters therefore reduce the emission of carbon monoxide, nitrates and organic hydrocarbons but actually increase the emission of CO_2. They reduce fuel efficiency and lead to more consumption. So whilst the universal fitting of catalytic converters will substantially reduce the emission of substances harmful to the environment, they will not, by themselves, solve all the problems.

One of the most intrusive effects of transport, particularly in urban areas, is noise. Most attention is paid to the noise nuisance created by road traffic and by aircraft, although rail operations can also be exceedingly noisy, particularly if they involve the use of high speed diesel units or passage over points.

Minimisation of the intrusive effects of aircraft noise depends on the move towards quieter aircraft, which is being encouraged by governments throughout the world, and by the control of aircraft movements, particularly during take off and landing. First generation jet aircraft may no longer be used for flights within Europe and European airlines are not permitted to add second generation jets to their existing fleets. All second generation jets should be phased out by 2002.

Most airports have rules governing the amount of night flying, and control the flightpaths adopted by pilots on take off and landing. Heathrow, Gatwick and Stansted are subject to rules restricting night flying by the noisiest types of aircraft laid down by the Department of Transport. Most other airports are subject to regulations laid down by their owners or planning authorities. These rules all tend to encourage operators to acquire quieter aircraft, and that is the direction in which things are moving. In some instances, however, operators and airport authorities are concerned that rules which are too rigid can threaten the viability of the airline and airport. This is particularly the case where the major part of the business is concerned with short-haul charter flights to the Mediterranean, which is the case in a lot of medium sized, regional airports. In order to get maximum use out of an aircraft during a relatively short season, it is necessary to make three return flights to a holiday destination in twenty-four hours, i.e. the aircraft must land and take off at its UK base every eight hours. If its turn round time on the ground is two hours, there can be no more than six hours between movements on the ground, at least one of which must come during a 'night' of 10 pm to 6 am.

The main method used for limiting the intrusive effects of road noise is to allow, as part of the cost of a new road, for the construction of noise barriers and double glazing of private property, under the terms of the Land Compensation Act 1973. Further details of the measurement of noise and the working of the regulations made under the Act are included in Chapter 19. The Transport Research Laboratory has a project in hand to determine the feasibility of manufacturing quieter heavy vehicles (QHV90). The quality of motorcycle silencing systems is controlled under the Motor Cycle Noise Act 1987. The Government is currently looking at the impact of car alarm systems with a view to minimising their intrusive aspects without compromising effectiveness.

Railway noise has been progressively reduced by improvements to the rolling stock and the gradual replacement of old track by continuously welded rail, thereby eliminating most of the joints which were the source of the traditional 'clackety-clack' associated with trains. The decline in wagon-load freight traffic and the marshalling of both passenger and goods trains into multi-vehicle units which are seldom broken down has reduced the amount of shunting dramatically.

Effect on towns, cities and the countryside

A less easy environmental impact of transport to measure is the effect that it has on the countryside and fabric of towns and cities.

Our planning system categorises land use, and identifies certain areas as National Parks, Areas of Outstanding Natural Beauty and Sites of Special Scientific Interest (SSSI). A decision to affect any of these categories by, for instance, the construction of a road, will require particular planning procedures and may render the scheme uneconomic or detract from its advantages as measured in purely operational terms. The result tends to be a compromise which, like many compromises, seems to please no-one. On the one hand, the environmentalist lobby would like to see all such sites as sacrosanct, whereas the roadbuilders complain that they must find the funds for diverting the road. There may be conflicts between one environmental interest and another. For instance, the A6 Market Harborough By-pass[2] has been routed to avoid a SSSI, but the result is more intrusive visually.

A lot of progress has been made in creating townscapes that are visually attractive and are no longer designed entirely around the needs of the motor car. Pedestrianisation of shopping streets and market squares has been a feature of urban planning for many years now, and the same ideas are beginning to be applied to residential areas. That this has been possible has been, to a large extent, the result of new road construction to provide by-passes and inner ring roads, or of the diversion of traffic away from shopping streets. It has therefore been bought at an environmental cost.

It would be unduly complacent to believe that we have now solved the environmental problems of our town centres. While congestion may have been removed from the High Street, it has, usually, been made worse on the immediate approaches to the town centre, and is going to go on getting worse as car ownership and dependence on road goods transport increases. The specific parts of this problem that can usually be describes as 'environmental' include damage to buildings due to vibration and exhaust fumes, damage to services, visual intrusion, danger to cyclists and pedestrians (real or imagined) and community severance.

European directives

One of the areas in which the European Commission has been most active has been environmental quality, and this has led to some conflict between the commission and the British Government, particularly in respect of the UK Roads Programme.

The dispute has arisen over the implementation of EEC Directive 85/337. The directive requires that an assessment must be undertaken to identify, describe and assess direct and indirect effects on:

- human beings, flora and fauna;
- soil, water, air, climate and landscape;
- interaction between the above;
- material assets and cultural heritage.

The assessment must:

1. describe the scheme and its site;
2. state the measures designed to mitigate adverse environmental impact;
3. give the data required to identify and assess the main effects which the project is likely to have on the environment;
4. contain a non-technical summary.

The developer must outline the alternatives considered and the reasons for his choice. He must describe the likely significant effects of the environment resulting from:

1. the existence of the project;
2. the use of natural resources;
3. the emission of pollutants;
4. creation of nuisance and elimination of waste.

Forecasting methods adopted must be described.

The Department of Transport has issued a Departmental Standard, HD18/88, to explain how the provisions of the directive are to be followed in the case of British trunk roads. As with other departmental standards, this can be treated as a definition of reasonable practice to be followed by other authorities.

The directive refers to 'express roads' as being mandatorily covered, and these are taken to mean 'special roads' within the meaning of the Highways Act 1980. The Government has also decided to include:

- all new roads over 10 km in length;
- all new roads over 1 km in length which pass through or within 100 m of a National Park, a Site of Special Scientific Interest, a conservation area or a nature reserve;

- all new roads over 1 km in length which pass through an urban area where 1500 or more dwellings lie within 100 m of the centreline of the new road;
- motorways or other road improvements which are likely to have a significant effect on the environment;
- other new roads likely to have a significant impact.

Within the preparation programme for trunk road schemes, the Department requires the preparation of an Environmental Statement after the selection of the preferred scheme but prior to any public inquiry. Where the environmental impacts of schemes other than the preferred scheme are signficiantly different, the department will include appropriate details under the reasons given for selecting the preferred scheme.

The requirements of the European directive are clearly open ended and open to interpretation. It is thus not entirely surprising that they have led to disputes arising between the British Government and the European Commission, and these disputes might yet have to be settled in the European Court.

The most publicised cases have been the extension of the M3 through Twyford Down, near Winchester in Hampshire, and the approach road to the proposed East London River Crossing at Oxleas Wood. In both cases, it is alleged that the construction of the new road would cause irreparable environmental damage. At Winchester, the proposal is to take the road through a deep cutting which would be visually damaging and would lead to the destruction of rare flora on the existing downland. An alternative route involving construction of tunnels was rejected on cost grounds. At Oxleas Wood, the road would involve the destruction of one of the very few areas of natural woodland remaining in the south east of England.

The Department of Transport has not prepared an Environmental Statement along the lines required by the Directive on the grounds that they were both in the preparation programme before the passage of the Directive, and therefore are not covered by it. This may be true in a strict legal sense, but it is hardly consistent with the Government's claim to be concerned about the environment. The damage will probably have been done long before the legalities can be tested through the courts.

Pressure groups

Because of the amount of public interest in environmental issues it is natural that groups should be formed in order to endeavour to promote particular policies. Many of these are now well organised and well funded and are an essential part of any constructive debate on transport issues. Most will form a view on environmental issues, even if the environment is not central to their objectives.

In contributing to the debate on transport, it is usually helpful, from everybody's point of view, if individuals of similar mind can combine into organised groups. They are likely to be taken more seriously by politicians, they can combine resources to undertake research or prepare evidence, they will attract the attention of the media and organisations are given precedence over individuals at public inquiries. The organisation of the group, of course, is infinitely variable, and its case can be wrecked by incompetent management. In general, however, it is the policy of governments in democratic countries to welcome participation by members of the public, and the higher the level of expertise on which that participation is based, the better. Some groups, such as English Heritage, are quasi-government bodies and look to Whitehall for funding. Others, like the National Trust, are actually independent but are recognised by government in various ways. Others, again, like the national Park Authorities, are established by central government but are really part of local government.

Engineers, therefore, should not be shy of involving themselves in pressure groups or giving advice, on a proper professional basis, to particular interests, although they will need to watch out for possible conflicts of interest, particularly if they are employed by public bodies.

Pressure groups can be divided into three categories:

1. Those which are government created organisations, which look to government for funding or other influence. The Conservative Government elected in 1979 vowed to sweep away these 'QUANGOs' (Quasi Government Autonomous National Organisations) although they did not actually get very far and there are as many now as there ever were. There are various stages in the preparation of highway schemes when it is necessary for the promoters to consult with such bodies.

2. National or international bodies which are essentially political, although they may not put forward candidates at elections. All political parties do come under this heading, although it is only the Green Party that could claim that environmental issues are its reason for existence. Bodies like Friends of the Earth, Greenpeace, Transport 2000, and the Environmental Transport Association are all genuinely independent organisations whose main purpose is to influence politicians and other decision makers. Other organisations contribute to or form lobby groups. Amongst these are the Road Haulage Association, Bus and Coach Council and the transport trade unions. There are a lot of other bodies which are not primarily there to influence transport policy but will speak up when the need arises: British Rail, local chambers of

commerce and industry, the motor manufacturers and the professional institutions.

3. Local bodies established to influence decisions on a particular scheme. These may be the most difficult to organise or to deal with as they will not have many resources, and frequently cannot afford either legal or technical representation. They are very prone to the NIMBY (Not In My Back Yard) syndrome, which is not very constructive and is hardly likely to appeal to inspectors at inquiries or elected members of planning committees. Part of the problem is that almost any transport proposal will be somebody's disadvantage. Perhaps the system would work better if less attention was paid to people's right to be heard and more to their right to be compensated.

References and further reading — Chapter 17

1. Waters M H L. *Road vehicle economy: state of the art review No 3.* Transport and Road Research Laboratory. HMSO. London. 1992.
2. Department of Transport. *Transport and the Environment.* HMSO. London. 1991.

Chapter 18

Environmental Impacts

Measurement of environmental impacts

In Chapter 17, we have seen that costs and benefits of transport investment need to be seen in environmental, as well as economic, terms. Some kind of trade-off is necessary if we are to reach logical decisions about investing money in items which cannot be easily measured in money terms. Cost benefit analysis, as applied through the Department of Transport appraisal programme, COBA,[1] considers everything in terms of monetary value and therefore does not include those other factors, globally referred to as environmental costs and benefits, which cannot be so considered.

The initial doubts on using COBA in isolation were voiced in the 1977 Report of the Advisory Committee on Trunk Road Assessment,[2] chaired by Sir George Leitch and frequently referred to as the Leitch Report.

The main recommendations of Leitch were:

1. that COBA should not be used in isolation, and that a framework should be developed, in addition to COBA, for scheme appraisal;
2. that the framework should take account of the needs of:
   (a) road users directly affected by the scheme
   (b) non road users directly affected by the scheme
   (c) those concerned with 'intrinsic value'

(d) those indirectly affected by the scheme

(e) the financing authority;

3. that the framework should be used as a basis of judgement between alternative options taking all factors into account, whether or not these could or should be expressed in monetary terms;

4. that a Standing Committee should be established to keep under review matters affecting the appraisal of trunk road schemes.

The then Secretary of State for Transport, William Rodgers accepted the principal recommendations of the report and established the Standing Committee referred to in recommendation (4), again to be chaired by Sir George Leitch. This became known as SACTRA (Standing Committee on Trunk Road Assessment) and has, to date, produced three reports, in 1979,[3] 1986[4] and 1992.[5]

The 1979 SACTRA Report set out the form that a framework should take and suggested that it should be prepared at three separate stages in the preparation of a trunk road scheme:

1. Public consultation stage.

2. Post-public consultation.

3. Public inquiry stage.

The recommendations of the report were substantially adopted by the Department and published in the form of Departmental Standard TD/12/83[6] and the Manual of Environmental Appraisal (MEA).[7] These identify only two stages at which a framework needs to be prepared and published: public consultation and public inquiry. Like other aspects of scheme preparation, however, it is constantly updated as the project moves forward.

TD12/83 defines a framework as 'a tabular presentation of data summarising the main direct and indirect impacts on people of the alternative options for a proposed highway scheme'. Thus, each option for the scheme identified as feasible by stage (7) of the twenty-four stages listed in Chapter 8 will now be represented by a column in the public consultation framework. In addition, a column of data will be prepared representing a 'do nothing' option, or, if not proceeding with the scheme would still lead to certain works becoming essential, a 'do minimum' option. The standard format for a framework requires a comments column.

The rows of data contained within the framework are grouped together in six groups, which are not quite the same as those recommended by the SACTRA 1979 Report:

Group 1 – the effects on travellers;

Group 2 – the effects on occupiers of property;

Group 3 – the effects on users of facilities;

Group 4 – the effects on policies for conserving and enhancing the area;

Group 5 – the effects on policies for development and transport;

Group 6 – financial effects.

Each of these groups may have sub-groups, and each sub-group will be subject to a range of impacts. A summary of the impacts to be considered when drawing up a framework is shown below:

Group 1

Time savings or delays expressed in monetary terms and discounted to the base year.

Changes in vehicle operating costs expressed in monetary terms and discounted to the base year.

Accident reductions expressed in monetary terms and discounted to the base year but supplemented by actual figures for fatal, serious, slight and damage only accidents.

Driver stress.

View from the road.

Amenity and severance effects for pedestrians and cyclists.

Group 2

Number of properties, broken down by sub-groups of residential, industrial and commercial, schools, hospitals, public buildings or other special buildings to be demolished.

Noise change, visual effects and severance for each sub-group of buildings.

Disruption during construction.

Hectares of land, recreational and agricultural, taken.

Group 3

Vehicle/pedestrian conflict, noise, visual intrusion and severance effects on sub-groups of shopping centres, public buildings and recreational facilities.

Group 4

Each policy of local and national government affected by the proposal will form a separate sub-group, e.g. the need to protect listed buildings, the need to protect conservation areas or proposals to develop a country

Group 5

Each policy of local or national government relevant to transport, development or economic policy will need to be listed here, e.g. the need to improve trunk roads in the area, the need to relieve through traffic from village and town centres, the need to develop a new industrial park.

Group 6

The monetary cost will be shown, divided by construction cost and land cost.

The monetary benefit, established via COBA, will be shown in two forms, one based on low growth traffic forecasts and the other based on high growth traffic forecasts.

The relative value of each proposal will be shown based on either a rate of return or the net present value of the option compared with the do-minimum option.

The impacts listed in the MEA for inclusion at various points in the framework are as follows:

Traffic Noise
Visual impact
Air pollution
Community severance
Effects on agriculture
Effects on conservation areas
Disruption during construction
Accidents View from the road
Driver stress.

The type of framework developed for the MEA and still in use, pending a major revision of the MEA, falls short of the original SACTRA recommendation by placing too much reliance on monetary values. These involve hidden factors and may favour particular options which, with different weighting, would not have been favoured. It assumes a relativity between, say, time savings and accident savings, which does not exist.

It can also be argued that several important impacts are left out simply because they are difficult to quantify. A section of road close to a school, for instance, may lead to parents delivering their children to the door rather than let them walk and risk crossing a busy road. Actual accidents show up in the framework. Inconvenience, which may prevent accidents, does not.

The second SACTRA Report, prepared under the chairmanship of Professor T. E. H. Williams of Southampton University, was specifically charged with looking at urban impacts. It largely confirmed the earlier proposals that decisions should be based on a combination of cost benefit analysis and judgement, and that the framework was a sensible aid to judgement. Points scoring systems were specifically excluded from the recommendations as it was difficult, if not impossible, to establish a satisfactory weighting system available for every case.

The 1986 report suggested that noise impact was particularly important in urban areas and a separate impact of night time noise should be included. The effect on conservation areas should be extended to include the effect on townscape, whether or not it is classified as a conservation area. The impacts headed 'view from the road' and 'driver stress' should be omitted. In appropriate cases, a statement regarding the social consequences of blight and development potential should be included. An alternative to the framework called the Assessment Summary Report was suggested. This had five sections:

1. Objections and problems
2. Opinions and consultations
3. Traffic appraisal
4. Economic evaluation
5. Environmental and social impacts.

Some experimentation has been undertaken in this area, but it has not been generally implemented.

The *Manual of Environmental Appraisal*

The *Manual of Environmental Appraisal* gives precise advice on the measurement of various impacts for inclusion in the framework. Some comments on each are set out below.

Traffic noise

One of the most apparent intrusions of road traffic is the noise of vehicles travelling along the road. The noise comes from the working of the engine, the friction between the tyres and the road, the movement of all moving parts of the vehicle, vibration of bodywork and components and the noise of brakes.

The unit of noise used is the decibel, which is related to pressure levels created on the eardrum by release of sound energy. The decibel scale is subject to weighting in order to emphasise those frequencies most readily detected by the human ear. The 'A' weighting has been found to give the best correlation between perceived loudness and pressure levels created. Traffic noise levels are constantly varying, so the normal measure used in Britain is the L_{10} (18 hour) level, that is, the noise level measured in dB(A) that is exceeded for 10% of the time through an 18 hour (0600–2400) day.

Noise levels can be measured directly using a noise meter, although it is important to note that readings can be affected by factors other than the loudness of the source noise, including weather and seasonal conditions; dB(A) is a logarithmic scale. Values approximating to certain common conditions are shown in the Manual of Environmental Appraisal as follows:

Threshold of hearing   0 dB(A)
A quiet bedroom   37 dB(A)

Communication becomes difficult	56 dB(A)
Busy office	60 dB(A)
Car 7 m away	71 dB(A)
Heavy diesel lorry at 40 km/h 7 m away	85 dB(A)
Hazard from continuous exposure	90 dB(A)
Pneumatic drill 7 m away	95 dB(A)
Jet aircraft 250 m overhead	102 dB(A)
Threshold of pain	120 dB(A)

The point at which noise levels are considered to be of sufficient nuisance to justify compensation is an L_{10} 18-hour level of 68 dB(A).

Compensation provisions are covered by the Noise Insulations Regulations 1975,[8] and normally take the form of installing double glazing in those houses deemed to come within the regulations. Three conditions have to be satisfied:

1. the combined expected maximum traffic noise level, i.e. the relevant noise level, from the new or altered highway together with other traffic in the vicinity must not be less than the specified noise level [68dB(A) L_{10}(18 hour)];
2. the relevant noise level is at least 1.0 dB(A) more than the prevailing noise level, i.e. the total traffic noise level existing before the works to construct or improve the highway were begun;
3. the contribution to the increase in the relevant noise level from the new or altered highway must be at least 1.0 dB(A).

It is clearly necessary to have some method of predicting the noise that will result from a new or improved road at the time that the detailed design of the road is being undertaken. The Department of Transport has laid down a standard method in the published booklet *Calculation of Road Traffic Noise.*[9]

In practice, the level of noise will vary significantly along the length of road under consideration, due to such changes as gradient, alignment and screening. Thus it is initially necessary to divide the length of road into a series of segments and treat each separately as a noise source. The intention should be to ensure that the variation within any segment should be no greater than 2 dB(A).

The predicted basic noise level for any segment is determined from a series of charts given in reference 8 and depends upon the traffic volume, traffic speed, traffic composition (proportion of heavy vehicles), the gradient and the road surface. Once the basic noise level has been determined, it is necessary to apply a series of corrections depending on distance, ground cover and whether propagation of the noise is obstructed or unobstructed.

Calculation of road traffic noise also specifies a full and shortened procedure for measurement of noise levels.

Although planning the location of new roads will take potential noise generation into account, along with other environmental effects, the primary means of minimising the nuisance is by the construction of barriers or increasing insulation of properties affected.

Barriers can take the form of timber fences, concrete walls or landscaped earth mounds. The noise reduction capability of barriers varies approximately with density:

for a reduction of 0 to 10 dB(A), M = 5kg/m^2
for a reduction of 10 to 15 dB(A), M = 10kg/m^2
for a reduction of 15 to 20 dB(A), M = 20kg/m^2

where M = mass (kg) per unit area (m^2) of the barrier material.[10]

Where additional insulation to buildings is provided, a rough value of the reduction in noise, also given in reference 9, is shown below:

Open single window	2–15 dB(A)
Shut single window	10–25 dB(A)
Shut double window (100 mm cavity)	25–38 dB(A)
Shut double window (50 mm cavity)	20–40 dB(A)

The other major source of noise of interest to transport planners and engineers is aircraft noise. Under the Civil Aviation Act 1982 the Secretary of State for Transport is empowered to impose noise reduction measures at specific airports. At Heathrow, the Department monitors thirteen sites around the airport and has a maximum level of 110 PNdB (Perceived Noise Level) for daytime flying and 102 PNdB for night-time flying. Operators are penalised for serious or repeated infringements, and there is thus an incentive on operators to fly their aircraft in such a way as to minimise the risk of breaking the rules.

Insulation of properties around airports is more difficult than those affected by road traffic, because there is less natural insulation between the house and the aircraft. All surfaces of the house, including the roof, are subject to noise penetration and the precise position of the source is less easy to pinpoint.

The generation of noise is a major factor in any planning decision concerning airports themselves or nearby developments. The unit used is the Noise and Number Index which is a function of the logarithmic average maximum noise levels in PNdB and the number of aircraft movements. NNI is normally expressed as an average value taken between 0700 and 1900 British Summer Time for the three month period from mid-June to mid-September. Points of equal NNI can then be joined by 'contours' to give a map of noise levels in the vicinity of an airport. Such contours should be drawn at 5 NNI intervals from 30 NNI upwards. A value of 35 NNI can be expected to give a low level of annoyance, 45 NNI moderate annoyance and 55 NNI and above serious annoyance. It is recommended that planning permission should not be granted for noise sensitive developments where the level is above 60 NNI and that care should be taken, and insulation provided, if the level is above 40 NNI.

These issues are unlikely to go away. With substantial increases in air travel predicted in the next twenty years or so, major airport expansion is expected. This has already seen Terminal 4 built at Heathrow, and pressure growing for Terminal 5. Stansted has been developed as London's third airport, a second terminal built at Gatwick and Luton has expanded dramatically. London City airport has come on stream in Docklands, while another development corporation, Sheffield, is building an entirely new airport which, like London City and Belfast City, will cater for STOL (Short Take Off and Landing) aircraft. Manchester has expanded, while a whole range of airports including Newcastle, Cardiff, Edinburgh, Glasgow, Birmingham, Leeds/Bradford, Teesside, and East Midlands have seen a steady growth in the number of passengers passing through. Even Liverpool, which is very much in the shadow of Manchester, has large scale plans for expansion involving reclaimed land in the Mersey Estuary. Closer European integration and the opportunities offered by the demise of the Eastern bloc can only lead to a greater demand for travel within Europe on both business and leisure, whilst increases in affluence would be likely to lead to greater demand for holidays in far flung places. As with other aspects of transport, demand for air travel is likely to grow with GDP.

Noise is a very subjective matter. While attempts have been made, in defining decibel scales, to correlate noise levels measured scientifically with human perceptions and annoyance, these can only be on the basis of average reactions. Degrees of annoyance are not only dependent on the characteristics of the individual being annoyed and the noise level being emitted: they can also depend on the tone of the noise and on the climatic conditions. Some would argue that a sudden, loud noise (e.g. an aircraft flying overhead at low level) is more annoying than a steady noise (e.g. road traffic) of the same intensity.

It is becoming more apparent that exposure to noise over a long period can be damaging to health. This is leading, for instance, to the adoption of Health and Safety Regulations concerning the wearing of ear protectors, with possible prosecution for the employer who does not provide them. But which is more likely to lead to long term health problems: using a pneumatic drill to open a hole in the road for, perhaps, two hours in a working day; or living under the flightpath of a busy international airport?

Transport planners are faced with a particularly difficult problem. Installations which generate a lot of noise, particularly airports, usually bring with them very substantial economic benefits. But it is almost impossible to apply precise measures to either the benefits or the costs. In such a situation, decisions must be made through the political process based on the best advice that planners and engineers are able to provide.

Visual impact

The MEA defines two kinds of visual impact: obstruction and intrusion. Obstruction occurs when the road, or some feature of it such as an embankment, a viaduct or a sign gantry, impedes a view that would otherwise be available. Quantification is based on the proportion of the view that is obstructed, measured by the solid angle subtended by an observer's vision when facing at right angles to the obstruction. It is classified on a three point scale as high, moderate or slight obstruction according to whether human responses are estimated as highly dissatisfied, dissatisfied or indifferent.

This approach involves trying to put numbers to something where no numbers are justified, and therefore is directly contrary to one of the main recommendations to the 1986 SACTRA Report. While it is probably safe to say that the construction of a motorway embankment at the bottom of a suburban garden would induce a high level of dissatisfaction, the same might not be true of a low level embankment at a reasonable distance in open country. Human responses are very subjective, and vary considerably from individual to individual.

Visual intrusion is essentially subjective, although some attempts can be made to classify the landscape through which a road passes. A special report by a qualified landscape architect may be required. Once again, the MEA calls for categorisation of visual intrusion into high, moderate or slight bands according to whether it would be expected to induce responses of highly dissatisfied, dissatisfied or indifferent. It requires some form of quantification of the number of people likely to be affected.

Assessment of visual intrusion creates the problem of determining in advance, what the landscape will look like after the road is built. It is now a relatively simple matter to compute a three dimensional model of the proposed road and create a perspective image, using readily available software. Photographic techniques can be used to produce a photomontage, i.e. a photograph or a particular view adapted to show the proposed road. This technique is widely used in public consultation and inquiry.

Air pollution

The amount of air pollution created by the new road will not normally vary significantly between alternative options for the same project, unless there are likely to be particular problem areas. It is unlikely that we would now wish to construct a major interchange close to residential areas in the way that Gravelly Hill on the M6 was constructed in Birmingham, leading to allegations of unacceptable levels of lead in the atmosphere on nearby housing estates. Tunnels will require significant measures to reduce the danger of a build up of noxious fumes in any case.

It is unlikely that air pollution will appear as a specific item in a framework, it being more appropriate to consider it as part of the global transport strategy developed by the government.

Community severance

The construction of a new road will impede journeys across the line of the road either because the journeys will be longer in distance and/or time or because travellers are dissuaded from making the journey at all. Severance may be absolute, i.e. caused by a physical barrier, or relative, caused by the difficulty of crossing a road carrying significant traffic. It is less likely to occur in rural areas and the demise of major urban road schemes makes the quantification of severance less relevant. However, the MEA recommends the inclusion of severance classified on a four point scale of none, slight, moderate or severe, according to the number of people affected and the degree of severance. For instance, an increase in the distance of pedestrian journeys by 250–500 m or of car journeys by 5–10 minutes would be classified as 'moderate', so would a pedestrian at-grade crossing of a new road carrying 8000–16000 vehicles per day.

It is pointed out that new road construction might result in a significant reduction in severance when traffic is transferred to the new road. A by-pass might lead to the pedestrianisation of a former High Street, resulting in the complete removal of severe severance for a large number of people.

Effects on agriculture

The loss of good quality agricultural land has sometimes been expressed as a concern by those opposed to the construction of new roads, although perceptions have changed somewhat with attempts to take land out of production as a means of reducing EEC food surpluses. However, if land is used to build a road on, it is not therefore available for any other use, and the relative amount of land taken for different schemes is a reasonable comparative statistic which may not be accurately reflected in the price paid for the land.

Land quality is graded on a scale from 1 to 5, 1 reflecting the best quality which has no limitations for agricultural use and 5 having very severe limitation due to soil quality and/or relief.

Agriculture may be affected if the line of the road crosses fields, dividing them into two parts, one or both of which is too small or wrongly shaped to allow efficient continued use. It might cause a barrier to the passage of livestock or farm vehicles. These factors should be taken into account when fixing the line of the proposed road.

Heritage and conservation areas

These fall into two groups:

1. Man-made structures, buildings and groups of buildings, which may be defined as ancient monuments, conservation areas, listed buildings, buildings subject to a preservation order or other historic sites.
2. Natural environments and habitats, or those originally created by man in which natural developments have taken place. These include the national parks, areas of outstanding natural beauty, nature reserves (whether publicly or privately owned) and wetlands, sites of special scientific interest, National Trust property, areas of outstanding landscape value, country parks, common land or church land.

Such sites are designated by either the local authority or the government under legislation relating to town and country planning, local government or access to the countryside, and lists are kept by local authorities and the Department of the Environment.

The designation of a particular building or location will not mean that the land cannot, in any circumstances, be used for road building, but it does mean that particular procedures have to be followed and, in many cases, it may be advisable to route the new road so as to avoid it. The framework will have to include specific mention of any of these categories.

Ecological factors

In addition to areas specifically considered above, there may be a large number of types of habitat which are not specifically protected but which are, nevertheless, important to wildlife and plant life. These may not be directly on the line of the new road but may be some way away but can be crucially affected by interuption of movement or drainage patterns. Sensitive areas might include:

1. rivers or streams
2. permanent pasture or unploughed meadows
3. areas of deciduous woodland
4. hedgerows
5. heathlands
6. bogs and marshes
7. mountains and open moorland
8. coastal areas, salting and estuaries
9. sand dunes.

An ecological map should be prepared to identify land under any of these headings and the impacts that the scheme will have. This may be a feature in determining the preferred route for the scheme or in some cases, such as the badger tunnels under the M4 in Berkshire, scheme design can be adjusted to minimise impacts.

Disruption due to construction

Some disruption, in the form of delays to traffic, noise and vibrations due to construction plant and the nuisance in the form of mud and dirt on the highway, is inevitable during the construction stage of the works, and a statement of the amount of such disruption should be included in the framework. It is important that contract conditions in the form of restricting times of operation, cleaning roads, provision of adequate signing and diversionary routes are rigorously enforced, and at the appraisal stage it is assumed that they will be.

This factor may be a major influence on the way in which a scheme is constructed, and possibly the way it is designed. Current motorway widening on the M5 south of Worcester is being done by constructing a new carriageway on one side of existing motorway, which can be undertaken without much interference with existing traffic flow. Once the new carriageway is completed, traffic in both directions will switch to it, and the second carriageway constructed on the line of the existing motorway. This means that the future centreline of the road will be offset from the existing, rather than retaining the same centreline and widening on both sides. The system means that all the overbridges will have to be completely rebuilt, but this is justified by the savings in delay to traffic using the road and the additional ease of operation for the contractor.

Pedestrians and cyclists

The effects on pedestrians and cyclists fall into two categories: journey time and amenity. Amenity is concerned with the pleasantness of the journey and this in turn is seen in terms of exposure to fear, noise, dirt and air pollution.

To some extent, these issues are covered under other areas incuding time savings and community severance. But they should also be considered separately where there are significant flows identifiable.

MEA recommends that the number of people affected by the proposed scheme should be identified wherever possible.

View from the road

The 1986 SACTRA Report recommended that this heading should be omitted from appraisals of urban schemes. Its use can also give some rather peculiar results in rural areas.

The basis of it is that the construction of a new road opens up views of the countryside to people in vehicles who would not, without the road, be there. This suggests that, for instance, the construction of the M6 through the Lune Gorge or the M62 over Windy Hill represent environmental 'gains'.

The fallacy arises from the assumption that the effects on 1000 people is always 1000 times the effect on one person. The justification for removal of this item from urban appraisals presumably follows from the fact that a driver has to concentrate for the whole of his time on other traffic and that passengers will have their view obstructed for most of the time. It might also be pointed out that the only view a driver gets whilst driving on the M6 or M62 is of the rear end of the vehicle in front.

Driver stress

The original framework suggested in the first SACTRA report included driver stress as a significant impact, although there are considerable difficulties in attempting to evaluate it as all drivers are different. The 1986 SACTRA Report recommended that it should be omitted in urban areas. Stress is thought to be made up of three components: frustration, fear of accidents and uncertainty regarding route to be followed. It increases as traffic flows, speeds and the proportion of heavy goods vehicles increase. It is presumably the intention of those who lay down current design standards that stress factors should be minimised, so stress should be low on newly constructed roads, particularly in the early years of their life. It may be, however, that elimination of stress is an important factor in justification of constructing any new road, i.e. in the comparison with a 'do minimum' option.

Present day practice

There have been substantial changes in public and political attitudes to the environment in the years since the Manual of Environmental Appraisal was published in 1983. The MEA itself is under revision to take account of the requirements of the European Commission outlined in Directive 85/337[11] and various UK regulations.[12,13,14]

Ecological effects of new roads have been summarised by Box and Forbes[15] as follows. Most of these effects would apply to other transportation developments such as railway lines.

1. Loss of natural features forming wildlife habitats, leading to changes in properties such as light, wind, temperature, humidity

and soil nutrients. If a habitat is split the population of each half is likely to be less than half the original population of the whole.

2. Hydrological changes to groundwater and surface run off. The construction of a new road may dam flow patterns leading to changes upstream, while changed run off times, acidity and the risk of pollution will lead to changes downstream. All these factors will seriously affect habitats.

3. Other impacts: the breeding success of certain species of bird has been found to be reduced within distances of up to 1.8 km of new roads. The road acts as a barrier to species trying to cross it, as evidenced by the number of dead animals found on the carriageway. Certain species will not attempt to cross the road, while others will use it as a relatively easy longitudinal route, leading to geographic spread.

4. Road led development. The corridors established for new roads, particularly close to junctions, are popular with developers of other facilities and this will lead to increased pressure for development. All such developments should be subject to environmental procedures under the terms of the EC Directive if, 'by nature of their size, nature or location they are likely to have a significant effect on the environment'.

It is important that designers should fully appreciate the ecological effects of their action and seek to minimise impacts from an early stage in the design. Methods by which this might be achieved include the following:

1. Fix the line of the road so as to avoid, as far as possible, sensitive areas, bearing in mind that such areas might be some distance from the line of the road.

2. Minimise the land take, by using viaducts and tunnels instead of cuttings and embankments. Consider steepening side slopes by use of crib walls or reinforced earth techniques.

3. Carefully design the drainage system. Incorporate balancing ponds if this will help disperse storm surges and allow settlement of particles.

4. Carefully design earthworks operations so as to avoid, if possible, net import or export of soil. Locate tips or borrow pits, including temporary ones, so as to minimise the impact on wildlife habitats. Design haul roads with the same care as the main route.

5. Allow natural regeneration where possible, using native species from local sources. Use of topsoil and imported grass seed should be avoided.

6. Time construction operations to avoid critical breeding times.

7. Include the source of aggregates for concrete and bituminous materials, as well as the source of unbound sub-bases and capping layers, in the environmental evaluation of the scheme as a whole.

8. Provide special facilities, e.g. badger tunnels, where these will be

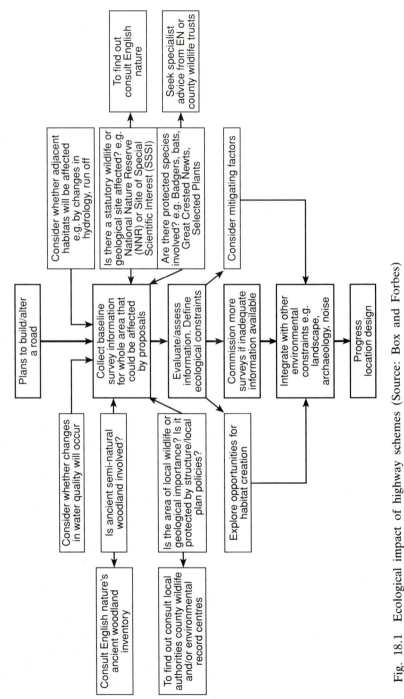

Fig. 18.1 Ecological impact of highway schemes (Source: Box and Forbes)

useful, and allow for the creation of habitats to replace those destroyed.

A line diagram showing the ecological input to highway works at the outline design stage is shown at Figure 18.1.

The British Government has been at pains to stress the value that it places on environmental issues, and this has been one reason for proposals to widen existing roads within existing corridors, as opposed to constructing new routes across greenfield sites, contained within the present roads programme.[16] The theme was continued with a more general environmental white paper[17] during Mr Chris Patten's tenure at the Department of the Environment. This is the background to the publication of the 1992 SACTRA Report,[5] produced by the Committee under the chairmanship of Mr D. A. Wood QC.

1992 SACTRA report

The Report starts from a specified position to the effect that a decision to build a new road depends on a number of considerations and constraints which sometimes conflict. These can sometimes be divided into 'economic' and 'environmental' factors.

Economic factors have traditionally been handled by use of the computer program, COBA.[1] This cannot cope with items that are not readily expressed in monetary terms, or, at least, have some common unit associated with them. In the early days of the appraisal of highway schemes, this fact was largely ignored. If something could not be readily expressed in terms of pounds and pennies, the only way in which it could affect the location or design of a highway was through the political process.

As has been stated above, this situation was officially recognised as being unsatisfactory with the publication and acceptance by the Government of the first Leitch Report (1977), which led to the adoption of the framework approach for assessing the impact of various alternative design solutions to a given scheme.

The 1992 SACTRA Report goes considerably further than earlier recommendations, and suggests that too much weight should not be given to the framework. There is a danger that it will lead to a mechanistic and oversimplified approach, and not enough attention will be shown to the fact that individual road schemes contribute to global impacts. The framework should not be included in the Environmental Statement required by the EC Directive, although some kind of matrix or checklist is clearly required if comparative statements are to be prepared.

The recommendations of the Report are:

1. Environmental assessment should be derived from specifically stated policy objectives. It must provide information on how the objectives are to be fulfilled by each option, including the do-minimum option.
2. Environmental assessments should ensure that every relevant environmental effect is included and is accurately described and/or measured.
3. Environmental assessment should be based on real evidence and the highest standards of technical information.
4. Resources used should not be out of proportion with the issues involved.
5. Consistency should be achieved across different alternatives and over time.
6. The procedures should be acceptable to the professional staff working on the project, and to the general public.
7. The language used should be clear and non-technical.

Individual schemes will contribute impacts on a national, regional and local scale, and this should be recognised in the preparation of Environmental Statements. The initial statement should be prepared at an early stage in the preparation programme, before the scheme is formally entered into the Roads Programme. Subsequent statements should be prepared and published at the public consultation stage, and at the stage when the line orders of the preferred route are published, prior to any public inquiry.

It is presumably the case that, at some point in the future, compatible systems of EIA will be adopted across the European Community. For the present, however, individual nations are left to develop their own legal requirements within the guidelines laid down by the EC Directive.

In studying a series of case studies in Ireland, Rogers[18] concludes that EIA techniques and methodologies are of central importance in the assessment of complex impacts on the environment.

The Irish Government has prepared guidelines[19] on the preparation of Environmental Impact Statements which derive directly from the EC Directive. These guidelines suggest four groups and effects on which to assess impacts, with units of measurement from monetary (time savings, vehicle operating costs, accident costs) to purely descriptive (landscape, culture). (See Table 18.1)

The output can be a framework not dissimilar to that used in Britain, a simple tabulation of data for comparison in which each alternative option is represented by a column of data, a checklist in which impacts can be represented by symbols representing positive or negative impacts at slight, noticeable or substantial levels or a checklist using a graphical ranking scale.

Table 18.1   Assessment groups and headings contained in Irish guidelines on EIA

Group	Effects
A. Road users	1. Time savings 2. Vehicle operating costs 3. Accident costs 4. Comfort and convenience 5. View from the road
B. Physical environment	6. Landscape 7. Infrastructure 8. Air quality 9. Nature conservation
C. Social environment	10. Community severance 11. Employment 12. Aesthetic 13. Culture, etc
D. Occupiers	14. Noise 15. Visual obstruction 16. Severance of land 17. Demolition

As in Britain, the comprehensiveness of the use of EIA techniques is increasing with time, and Rogers points to the two most recent studies considered, Northern Cross Motorway in Co. Dublin[20] and Kilmacanogue to Glen of the Downs improvement in Co. Wicklow[21] as representing the state of the art as laid down in the governmment's guidelines.

References and further reading – Chapter 18

1. Department of Transport. *COBA9 – A program for analysing costs and benefits of highway investment.* 1981. (As amended to 1990.)
2. Leitch, Sir George (Chairman). *The Report of the Advisory Committee on Trunk Road Assessment.* HMSO. 1977.
3. Leitch, Sir George (Chairman). *Trunk Road Proposals: a comprehensive framework for appraisal.* 1st Report of the Standing Committee for Trunk Road Assessment (SACTRA). HMSO. London. 1979.
4. Williams T E H (Chairman). *Urban Road Appraisal.* 2nd Report of SACTRA. HMSO. London. 1986.

5. Wood D A (Chairman). *Assessing the Environmental Impact of Road Schemes.* 3rd Report of SACTRA. HMSO. London. 1992.

6. Department of Transport. Departmental Standard TD12/83: *Frameworks for trunk road appraisal.* Department of Transport. London. 1983.

7. Department of Transport. *Manual for Environmental Appraisal.* London. 1983. (Under revision).

8. Statutory Instrument SI 1975 No 1763. *The Noise Insulation Regulations.* HMSO. 1975.

9. Department of Transport. *Calculation of road traffic noise.* London. HMSO. 1988.

10. Freeborn P T and Turner S W. *Environmental Noise and Vibration.* in Roberts J and Fairhall D (eds). *Noise Control in the Built Environment.* Gower Technical. Aldershot, England. 1988.

11. Commission of the European Communities. *The assessment of the effects of certain public and private projects on the environment.* Council Directive 85/337. Brussels, 1985.

12. *The Town and Country Planning (Assessment of Environmental Effects) Regulations 1988.* SI 1988 No 1199. HMSO, London. 1988.

13. *The Highways (Assessment of Environmental Effects) Regulations 1988.* SI 1988 No 1241. HMSO, London. 1988.

14. Department of the Environment. *Environmental Assessment.* Circular 23/88. HMSO, London. 1988.

15. Box J D and Forbes J E. *Ecological considerations in the environmental effect of road proposals.* Highways and Transportation. 39. 4: 16–22. April 1992.

16. Roads for prosperity: the Government's roads programme. HMSO, London. May 1989.

17. Cmnd 1200. *This Common Inheritance.* HMSO, London. September 1990.

18. Rogers D. *The use of EIA techniques and methodology within the road planning system in the Republic of Ireland.* Proc. Inst. Civ. Engs. Municipal Engineer. 93. 3: 39–50. March 1992.

19. McIlraith D. *Environmental impact assessment, roads: notes for the assistance of road authorities (draft).* Department of the Environment of Ireland. Dublin. 1990.

20. Dublin County Council. *Northern cross route motorway scheme: environmental impact study.* Dublin County Council. 1989.

21. Ove Arup and Partners. *Proposed Kilmacanogue to Glen of the Downs dual carriageway: environmental impact statement.* Wicklow County Council. June 1990.

Chapter 19

Minimisation of Environmental Impacts

Use of tunnels

There have not been many cases in Britain to date of changes in the design of schemes specifically to alleviate environmental problems.

It is sometimes pointed out that putting transport corridors in tunnels is an effective solution, although it will almost always lead to substantial additional costs. About the only British example of a road built in cut and cover tunnel when a cutting would have been a feasible option is the M25 between Junctions 26 and 27 where it crosses the ridge of Epping Forest. Other road tunnels are mostly concerned with estuarial crossings (Dartford, Blackwall, Conwy, Tyne, Mersey) and have been constructed by boring or, in the case of Conwy, immersed tube.

Cut and cover construction on the A1(M) at Hatfield and A58(M) in Leeds has primarily been to make high quality, high cost development land available over the top of the tunnel, although in the case of Leeds, the proximity of Leeds General Infirmary was a factor. The Leeds tunnels are operationally problematical, as they consist of dual two-lane carriageways without hard shoulders and have a central wall between the carriageways. They form part of the city's Inner Ring Road and carry heavy traffic. An accident or breakdown leads to severe congestion over a wide area and emergency vehicles are unable to reach the blockage. Cut and cover tunnels on the A658 between Bradford and Harrogate and the A538 between Wilmslow and Altrincham were built to allow runway extensions at Leeds/Bradford and Manchester Airports respectively.

Bored tunnels on the A26 at Lewes in East Sussex and A55 North Wales Coast Road between Conwy and Penmaenmawr in Gwynedd were constructed as the best way of dealing with the geological conditions in the locality, although both generate significant environmental benefits. Those on the A449 at Monmouth and M4 at Newport, Gwent were the shortest route between two points on opposite sides of a spur and were environmentally beneficial although, in Newport's case, operationally difficult.

The suggestion to build a tunnel to carry the M3 through Twyford Down near Winchester in Hampshire has been specifically rejected by the Government. Christopher Chope MP, then Minister for Roads and Traffic at the Department of Transport, has written:

> Some people have suggested that we build a tunnel so as to preserve the outward appearance of Twyford Down. This option was considered at the 1985 public inquiry and I recently considered a proposal from the Twyford Down Association for a toll tunnel.
>
> These have been rejected not only because of the increased cost, estimated at £90 million, and the delay to the scheme, a minimum of six years, but because they would also have a damaging effect on the environment. Two tunnels, each large enough to take three lanes of motorway traffic and a hard shoulder, would have prominent entrance portals in the landscape on either side of Twyford Down and would need to be lit, and there is spoil to get rid of.
>
> Charging tolls would not make the tunnel any more attractive, in fact it would make it less so. People who want to avoid paying the tolls would have to use smaller, unsuitable roads to by-pass the tunnel and may end up driving though Winchester itself. The only free, suitable alternative, would be the existing A33 Winchester by-pass which would have to remain, thereby removing the major environmental benefit of building the road through Twyford Down and restoring the link between the Water Meadows and St Catherine's Hill.[1]

Underground railways exist in most major cities, either as part of the urban transport system or to carry main lines into city centre termini. Many of these were built a hundred or more years ago on account of the need to bring the line through densely populated areas where land could not be made available on the surface. Modern projects, however, like the Victoria and Jubilee lines and Bank extension on the Docklands Light Railway in London do have environmental advantages. Many urban strategy studies concentrate on public transport not only because it makes the most efficient use of scarce land and hence reduces congestion, but because it is, comparatively, environmentally friendly.

Changing the alignment

There have been cases where the proposed line of a new road has been amended in order to avoid environmentally damaging areas, such as the M40 at Otmoor and Birmingham Northern Relief Road at Chasewater Heath. There is a possibility that the Department of Transport will abandon its proposals to build a motorway link between the M1 at Wooley and M62 at Brighouse, West Yorkshire, because of the extent of local opposition, a lot of which was centred around the fact that both alternative routes included at public consultation stage would be almost entirely on green belt land.

A major argument advanced by the Government in favour of its strategy of concentrating traffic on existing corridors rather than creating new ones is that many of the environmental impacts can thereby be avoided. Thus, the Government is pressing ahead as fast as it can with the widening of the A1 to full motorway standards between London and Newcastle-upon-Tyne, whilst it is giving very little encouragement to those who would create a new 'East Coast Motorway', extending the M11 from its present terminus at Cambridge via the Humber Bridge to Teesside.

Noise barriers

In urban areas, the construction of concrete fences can significantly reduce the noise impact of major roads, although it is likely they will cause some visual obstruction, and they hardly add to the quality of the view, either from on the road, or off it.

Where land is available, careful landscaping using surplus soil and planting areas can both reduce noise and enhance the visual amenity of the area.

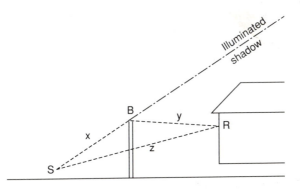

Path difference = x + y − z

Fig. 19.1 Noise barriers

References and further reading — Chapter 19

1. Chope, Christopher. *Ministerial statement published in Highways and Transportation*. 39. 3: 30. March 1992.

Chapter 20

The Environmental Dilemma

To conserve or to advance?

There are very few things that a transportation engineer can do which does not have some detrimental effect on the environment. Move a single piece of earth and you destroy the habitat of something. Use any kind of vehicle and you consume non-renewable resources, create noise and vibrations and run a risk of injury to yourself or others. If we give absolute values to environmental protection we will rapidly find that we can do nothing at all.

On the other hand, there is something of a consensus around the view that development in Britain and other advanced countries has often been at considerable environmental cost which has seldom been recognised, still less evaluated. Environmental impacts will be a more significant part of an engineer's life in the future than they have been in the past.

We are faced with the problem of feeding environmental issues into the decision making process in a way that will lead to rational, positive outputs. Decisions in Britain and other Western democracies are made by elected representatives on the basis of advice from professional officials. The purpose of legislation passed in the past ten years or so has been to formulate professional advice in a manner that will cause forecasts of environmental impact to be specifically and prominently stated alongside economic costs and benefits of a particular scheme. We then trust the elected representatives to make the right decision as far as trade-offs between different impacts are concerned. But who is to say that a nature reserve should or should not be destroyed in order to move traffic away from a village, where noise from passing trucks makes life intolerable for residents and there is a high risk of accident to young children on their way from school?

In the end, elected representatives make the decision. But elected representatives are politicians, subject to pressures from many different directions and concerned to implement their own ideology. There is a connection between the politician's decision and what he perceives to be the welfare of the community as a whole, but it is rather tenuous. And it is a matter of opinion as to whether there is a connection between the politicians' perception and reality. Politicians, by the nature of the beast, must be concerned with the people who elected them and their chances of re-election. That may give them a parochial outlook and the wish to pander to the wishes of a particular group which may be significant in their constituency. It will certainly emphasise the short-term nature of their office and a desire to be involved in activities that lead to a rapid and visible return. The environment, except in very localised and low-key situations, does not lend itself to such treatment.

There is often a conflict between environmental and economic objectives. We have been brought up to expect continual growth in the economy and the society we have in Britain today will not function without it. We have placed a high priority on obtaining consumer goods, on guaranteeing high quality health care and education and giving pensions to all when they have completed their working lives. We have often borrowed money to pay for all these things and we have set up a complex system of training and employment based around them. Changing tack and going without is not an option. Yet a transport system that has been developed with little concern for preserving the environment is often an essential ingredient to our way of life. We will have to consider what we are prepared to give up if we are to face up to local, national and global environmental issues and take proper account of the cost of destruction.

The engineer's contribution

Engineers can make an important contribution, particularly in the field of transportation.

For twenty years, now, engineers have recognised that it is uneconomic to construct a road for the maximum amount of traffic that will ever use it. Smaller design values have been used. Economic judgement has come to be used alongside engineering judgement in road design. The costs and benefits, expressed in pounds sterling, of many new roads have been evaluated, without much thought being given to the weighting factors implied in the calculation. Many issues, such as landscape and concentration of exhaust gasses and loss of agricultural land tended to be ignored for want of a method of placing a value on them.

From the original ACTRA report, the situation was recognised as being unsatisfactory and biased in favour of the presumption that roads would have to be built. The European Commission became involved, and was able to do so precisely because it is not subject to re-elections through the democratic process. Now we have a requirement for very specific statements on environmental impact, but they do not solve the basic dilemma. How do you trade off one man's gain against a loss which, arguably accrues, to everybody.

The engineer's life is becoming more difficult. We can no longer assume that our activities are subject to universal acclaim. We are likely to find ourselves, increasingly, on the dividing line between opposing factions. We do not have the right to resolve the dilemma, but because of our training and experience, we have the opportunity to advise on how it can be solved. We will have to become more proactive in environmental issues, more prepared to listen, more prepared to compromise. We will have to develop a good understanding of the political processes which are the only mechanism in our society for making difficult decisions. And we will have to keep an eye on the long term, because the politicans and the economists are primarily concerned with the short.

PART V

TRAFFIC MANAGEMENT

Chapter 21

Objectives of Traffic Management

Economy

Transport is, by its nature, located in both the private and public sectors. The private sector's objectives are easy to define. A company must, by law, seek to maximise the return on investment of its shareholders. It must, therefore, continually seek opportunities in the market whereby it can utilise its capital and secure its profit. It is not for a private company to seek to regulate or interfere with the market by, for instance, imposing traffic management measures. The company must accept the legal constraints upon its activities, but that is all.

The initiation of a traffic management scheme will almost certainly lie in the public domain. With few exceptions, highways are public objects and traffic management is a means of controlling the highway in the interests of the authority. There will be plenty of private involvement in the scheme – suppliers, contractors and maybe designers – but the scheme itself started off, and will end up, as public property. So the objectives of traffic management are likely to be defined by an engineer acting on behalf of a highway authority.

In the 1950s and 1960s, when the science of traffic management first became relevant to the work of the civil engineer, the objective could be very simply defined: it was to get the maximum amount of traffic through a given amount of road space. Efficiency was defined in terms of the amount of movement that was taking place, without reference to the benefits that could be gained as a result of that movement. The amount of investment that was available to improve the road system was limited more by legal constraints and shortage of manpower than it was by lack of financial resources. Public transport was seen as a necessary provision but one which would gradually diminish in importance until it disappeared altogether. The effect of transport on the environment was simply not appreciated or considered to be important.

Today, public sector investment has to be justified by a careful evaluation of the costs and benefits of any scheme, and this is as true for traffic management as for new road construction.

Control over the total amount of capital that is spent throughout the public sector is rigorously exercised by central government and the Treasury, whose objectives may be quite different from those of the engineer seeking to justify his scheme. Treasury objectives must be concerned with the overall management of the national economy. Restricting public sector expenditure in the interests of keeping inflation and interest rates down and encouraging higher levels of employment therefore have a high priority. This has had a serious effect on the promotion of traffic management schemes, although not as serious as on major new road building. Local authorities have seen the resources available to them dwindle each year. A major county authority, which might have had twenty or thirty important highway improvement schemes in preparation at the end of the 1960s, might now consider itself lucky if it has a single one. Even the Department of Transport's trunk roads programme looks less impressive when compared with the situation in the 1970s when hundreds of miles of motorway were coming on stream every year. A private company will continually seek to increase its investment levels and expand its activities, because that is the way in which it will be perceived as being successful. Public organisations, however, might find themselves expected to do the reverse.

One advantage of a traffic management scheme is that, by comparison with new construction, it is very cheap. Spending £100,000 may be able to achieve major changes in the way a transport system operates. The same money spent on new road construction might achieve a few metres of motorway.

But if their low cost makes traffic schemes attractive, it also introduces dangers. Many of the problems that we face today, particularly in public transport, arise from lack of investment in the past. Although much of the rail system is now blessed with new rolling stock this only came about when the old diesel multiple units were more than thirty years old. Now that many authorities have a specific policy of attracting people onto public transport as a means of improving the efficiency of land use, we might be entitled to wonder how many passengers have been frightened off in the past by dirty vehicles and dark and dingy terminals.

Efficiency

It is the efficiency with which capital is invested, rather than the size of the investment, that is important, both in justifying expenditure to central government and in maximising the benefit to the local community. But efficiency is not an easy thing to measure. The Government tends to view efficiency as being synonymous with value for money, but this reflects the importance placed on monetary value by a government of a particular political persuasion. A more thorough look at the type of costs and benefits that are implicit in transport is needed if we are approach a precise definition of efficiency.

A journey may be undertaken in the most efficient manner, but that does not add to the overall efficiency of the system if the journey itself was not needed or represented an inefficient way of achieving a primary objective. By and large, we have tended to regard transport as being a good thing because it is the means by which choice is made available: choice of goods to buy in the shops, choice of places to live, choice of jobs to do. But choice is restricted in many other ways, particularly through the mechanism of price, which may itself be due in part to transport. It may well be that the average commuter does not have much choice about where he is going to live. He buys a house at the nearest location to his place of work that he can afford. The availability of a highly efficient, cost-effective transport system has not made him better off. It has simply pushed up the price of houses in all the more convenient locations so that he can no longer afford them.

The same situation arises in goods transport. One side effect of the construction of the motorway system has been to cause large companies to locate warehouses close to the network. A company might well find that, by so doing, it can cover the whole country with say, four depots. It could cover the south west and South Wales from near Bristol; the north west and North Wales from near Warrington; the north east from near Leeds; and the south east and East Anglia from near Luton. Any part of England and Wales would be within daily travel for a delivery van. The journeys involved would be much longer than when the company operated out of fifteen separate depots, all convenient for railheads, but distance is nothing like as important as time for most companies. A lot more travel is taking place, and, no doubt, some of the benefits will be passed on to customers in the form of lower prices. But can we say that the system as a whole is more efficient, bearing in mind other factors like the job losses at the closed depots, the additional use of scarce fuel resources and contamination of the environment close to the motorways being used?

Environment

A consideration of efficiency in transport terms inevitably leads to a consideration of the environment. An individual's decision to travel (and a company's decision to locate a warehouse) will be based on that individual's perception of the benefits he will gain from the journey(s) and the actual cost that he will be required to pay. That cost will be measured in pounds. On public transport, it will be represented by the cost of the ticket. In a private car, it will include the cost of petrol that has to be bought, parking and probably some rather inexact element for the fixed costs of depreciation, insurance, road tax and maintenance. It will not include an evaluation of the cost to society as a whole of using up non-renewable resources or of passing large quantities of noxious fumes into the atmosphere. It is this inability of our economic system to take proper account of such things as environmental pollution that has led to the present level of concern for environmental issues.

A particular aspect of environmental concern that should be considered is safety. Under the present system of evaluating highway improvement and traffic and management schemes, reductions in the level of accidents are taken into account in COBA and in the preparation of a framework. For low cost schemes not subject to full evaluation, a three year accident history for the area to be subject to improvement will be included in the reports justifying expenditure. Accidents will be divided into three categories according to the most serious injury sustained: fatal, serious or slight. Damage only accidents will be largely ignored.

A monetary value will be placed on each category of injury, which will be intended to represent the cost in terms of emergency and medical treatment, police time involved, the loss of potential production of the person injured, and a 'subjective allowance' for pain and suffering that results from the accident.

The deficiencies of this approach in defining objectives in terms of maximising return on investment can be easily demonstrated. Subjectively, most people would agree that a serious injury accident is preferable to a fatal one. The 'subjective allowance' part of the calculation of costs as determined by the Department of Transport ensures that this is what the figures say. But if we are really concerned to minimise the costs to society as a whole, this conclusion is not logical. The worst thing that can happen, financially, is that a relatively young person is so seriously injured that they are prevented from working ever again. That, as insurance companies would agree, is an enormously expensive outcome. By comparison, the death of somebody towards the end of their working life when their dependants have become self-sufficient might actually result in a saving to the Exchequer, because they won't need a pension or the increasingly expensive medical facilities that are required into old age.

In evaluating the costs and benefits of traffic management schemes, the three headings of economy, efficiency and environment need to be considered separately.

Economy is important because there is a serious shortfall in the amount of resources that are available for investment in transport. We cannot afford to waste them. But the cause of economy should not be taken too far. Under investment in public sector projects has left us with serious problems which we still have to resolve.

Efficiency is difficult to define. Decisions are taken by individuals on the basis of what they perceive to be an efficient basis. But their perception may not be very accurate, and an inevitable consequence of one individual maximising his efficiency may be that another individual loses out.

The environment, including safety, is close to the top of present day policies. Procedures such as the EC requirement for environmental statements at various stages in a project's development, and the Government's stated policy of reducing the casualty rate from road accidents, suggest that we are getting nearer to identifying and setting down environmental impacts. But, beyond a general commitment to avoid general destruction, we are little further forward in defining objectives.

Whose cost? whose benefit?

It is fundamental to the present procedures that costs and benefits are considered to be those which accrue to society as a whole, and that these are considered to be the sum of all the costs and benefits that accrue to individuals added to those which accrue to the state.

Historically, when the priority has been in building major roads, the gainers have usually outnumbered the losers by a considerable margin; 100,000 people might travel along a length of motorway every day, and each one of them will experience some measure of gain in terms of reduced journey times, lower costs and lower risk of injury. The major part of the cost will have been borne by the state in the form of construction and maintenance costs. The number of people bearing costs individually in the form of lost amenity or noise intrusion, will be quite small. If 50 properties are affected by a kilometre of new road, there will be something of an outcry. The means of compensating the owners of those properties may be unsatisfactory and we may not have too much confidence that justice has been done, but it will clearly be the majority that has had its way.

With traffic management schemes, at least of the sort that are now coming on stream, the reverse may be true. Introducing traffic calming, particularly if it extends to main roads, will only be of direct benefit to those people who live or carry on business in the area immediately involved. Through traffic, which is likely to be the majority, will have to bear the cost in terms of slower journey times, and the state will still have to pay for construction and maintenance. Something of a change of philosophy seems to have taken place. With a major new section of motorway, a small number of people have to pay a large amount each in order that a large number of people can benefit by a small amount. Introducing traffic calming is likely to mean that a large number of people pay a small amount each in order that the few can benefit considerably.

This change has major implications for the way in which we approach a wide range of transport issues. It could be used, for instance, to revolutionise our approach to public transport in a way that would be anathema to the present Conservative Government. Take, for instance, the traffic-related problems experienced in the Lake District National Park, where enormous numbers of visitors arrive by car, clogging up the roads within the Park and for a long way beyond. Public transport, where it exists at all, is slow, inconvenient and expensive. Maybe it is possible to suggest that, if the people who now travel by car travelled by bus and train, the network would be able to cope and there would be no traffic jams. That is not the point. The point is that the people who travel by car have no wish to travel by bus or train, and they are clearly the majority. So we are left with an anarchic situation, in which chaos will continue to impose environmental and social costs until the situation becomes intolerable and draconian measures are taken to ban cars from the Lake District.

Would it not be better to change our definition of 'benefitting society as a whole' to something more complicated and far-sighted than simply summing up all the costs and benefits to individuals?

Chapter 22

Implementing Traffic Management

Movement control

A wide range of controls have been imposed on traffic movement, particularly in urban areas, for a long time. The freedom to travel upon the public highway has been qualified, at least since the eighteenth century, by legal measures. Initially, these were designed either to raise revenue or to ensure public safety. With the rapid growth of motor traffic after the Second World War, measures such as parking controls and one-way streets were seen as a means of facilitating the movement of through traffic. In the past twenty years, traffic signals, which were originally introduced in the nineteenth century as a means of eliminating conflict between crossing traffic streams, have become the major instrument of area wide traffic management.

Traffic signal installations fall into three groups:

- Stand alone signals at isolated junctions, controlled by a signal controller housed in a pillar close to the junction. These will be either vehicle actuated or operate on a fixed, preset time sequence and the timing will be determined for that junction alone, irrespective of what is happening elsewhere on the network.
- Linked signals, where two or more junctions are controlled by a single controller housed in a pillar close to the junctions. These will allow for a fixed, preset time sequence on all the signals

determined to favour the major flow, but otherwise irrespective of what is happening elsewhere on the network.

- Area wide systems, in which all the signals within a defined area are linked to a central computer which then determines individual settings either through a series of fixed plans (TRANSYT) or through a system which allows some degree of response to traffic levels as they occur (SCOOT). These systems, sometimes referred to as Urban Traffic Control (UTC), will normally be programmed to minimise total delay over the whole of the network.

The principles behind each of these three types of installation are described in Chapters 12 and 13, along with details of the calculations necessary to determine the optimum signal settings. Area wide systems are now seen as essential in urban centres if congestion is not to build up to an intolerable degree and the danger of locking a whole system is to be avoided. Because of the comparatively low cost of installation and the very high returns in terms of time savings comparatively small urban areas are adopting UTC systems. Central traffic control can be applied to tidal flow systems, allocating a greater proportion of road space to the dominant flow in morning and evening peaks; giving priority to buses and, in the near future, trams; combining fixed time plans with vehicle actuation on lightly used links or pedestrian crossings.

With the opening of the on-street running of trams in Manchester in the summer of 1992, new types of traffic signal control are required to ensure safe operation.

The main differences between a Light Rail Vehicle (LRV) and a normal road vehicle are:

1. Its weight, and therefore stopping distance, is greater.
2. It is limited to a fixed track and therefore cannot manoeuvre.
3. Its arrival will be fixed by a timetable which may mean less arrivals than one LRV per cycle.

Since LRT (Light Rapid Transit) systems are dependent upon attracting custom that would otherwise travel by car, LRVs should perhaps be given priority. The Manchester system is fully described in a paper by Tyson.[1]

Modern experience of installation of LRT systems with an element of on-street running is limited, but transportation consultants MVA have developed a program, FLEXSYT, to model signals so as to allow for on-street running. It has been possible to compare actual operation with the model in Hong Kong (Tuen Men LRT system).[2] Part of the system proposed for Croydon has also been modelled.[3]

The objective in Hong Kong is to give priority to LRVs whilst maintaining safe operation. The system opened in 1988. There were a number of accidents in the early days but the level has reduced as drivers became more familiar with the system.

Control of the signals is on a demand actuated system, with the request detector located on the track sufficiently far in advance for the light to turn to 'Proceed' before the LRV passes the critical braking point. The system must cope with a queue of LRVs forming and with a situation in which two or more LRVs approach from different directions at the same instant. Because of the size and acceleration and braking properties of the LRV longer intergreens are likely to be required than with conventional road traffic.

Sophisticated systems enable faults and other problems to be diagnosed quickly and action taken. If the problem is a technical malfunction, it is possible to get a repair crew on site rapidly, but there is still a need to inform drivers of the problem sufficiently far in advance to allow the driver to select an alternative route or at least avoid ploughing into stationary traffic ahead of him that he was not expecting to be there. This is part of a new approach known as demand management.

Demand management

Limited information can be put over to drivers using Variable Message Signs (VMS). Motorways in the UK have been equipped with VMS equipment, originally very rudimentary, for many years. It is now available for use in urban areas to show which car parks have space available. Information on whether a particular car park is full can be relayed to a driver some distance away so he can be directed to those car parks which have spaces. This is not only convenient for the driver, but saves a lot of unnecessary traffic on central streets caused by drivers hunting for a place to park. It is only a short step from this application of VMS to apply it to general route signposting. A particular route through a town might usually be the most satisfactory, so that is the route identified by the fixed direction signs. There might, however, be occasions, when that route becomes congested. Maybe it goes past a premier league football ground, so it gets very busy on a Saturday afternoon. VMS could be brought into use only on Saturday afternoons, and ensure that non-football traffic was kept away from the ground.

Local radio broadcasts are another method of communication between traffic control centres and drivers. Anyone who is used to driving around urban areas listening to the local radio station will be well aware of the shortcomings of this approach. Traffic congestion can build up and disperse within time periods of less than half an hour, which is likely to be a shorter period than the lead time for getting information broadcast. Even planned events, like contraflows on motorways, may be in place for several weeks but the degree of congestion caused varies by the hour.

The areas covered by local radio stations do not necessarily reflect movement patterns of travellers. Particular wavelengths are determined by geographical features in the landscape and the location of transmitters. BBC Radio Leeds has two VHF wavelengths, depending whether you are in the north or south of the county. It disappears altogether on the northern outskirts of Sheffield or when you drop down the western side of the Pennines heading for Manchester. This is not surprising given the shape of the land mass and the area the station serves, but it does mean that a motorist will have to retune several times in the course of quite a short journey. This has been recognised by the Department of Transport who have erected signs showing the local radio wavelengths along the motorways, but retuning the receiver in heavy traffic is not necessarily a very safe operation.

Where roadworks are programmed well in advance, a number of techniques can be used to keep drivers informed. Motoring organisations maintain information on recorded tapes which motorists can dial into by telephone. The Department of Transport distributes printed leaflets through service centres and garages. Television screens are now available in most motorway service areas giving out up to date information on roadworks, traffic conditions and the weather, and subscribers can get access to the same information through the CEEFAX, ORACLE and PRESTEL databases.

Very much more sophisticated approaches to demand management are currently under investigation. Systems are being developed whereby traffic information can be relayed to drivers using in-car receivers which will interrupt whatever programme is tuned in. Access to the system could be provided directly for the police so information could be broadcast without the delay implicit in going through a radio station. It would no longer need to matter whether the driver was tuned to the right station.

Management of traffic demand is not limited to providing information to travellers or potential travellers. Policies can be developed in an integrated planning system that will favour particular trips or trips made at a particular time. This already operates to some extent through, for instance, the mechanism of parking charges and the provision of off-peak fares on public transport. Whilst off-peak fares are clearly sensible from the transport operators point of view, because they tend to spread the load away from the peak and fill seats that would otherwise be empty, they may not be sensible from an integrated transport planning point of view. Congestion tends to occur during the peak hour, so the greatest benefits overall would be achieved by securing a modal shift during the peak, not the off-peak.

There may be scope for regulating road improvement schemes from the standpoint of affecting travel demand. By and large, a new road will encourage some people to make journeys that they would not otherwise have made. The question therefore needs to be asked as to what costs these new journeys will impose elsewhere on the network:

will they lead to additional demand at points where there is spare capacity, or at points where congestion is already building up?

The most direct method of limiting travel demand is through the pricing system, and it is now technologically possible to operate a road pricing system, whereby motorists are charged directly for use of the road. It involves the introduction of infrastructure in the form of cables and electronic gates, but it is considered to be feasible in city centres and certain highly trafficked toll routes.

Several systems are available but they all involve an electronic counter mounted on the vehicle and defined collection points through which the vehicle must pass if it is to gain access to the congested network. The most immediate application is for the automatic collection of tolls on busy roads, such as the Dartford river crossing. Regular drivers would be able to mount a transmitter or receiver on their vehicle which would communicate with the complementary device at the toll station. It could work on the basis of a prepayment smartcard mounted on the vehicle, or could be coded so that the vehicle was identified by the tolling station and an invoice subsequently despatched. The vehicle can pass through the tolling station without stopping, leading to a reduction in queues and delays.

The same principle could be used to levy charges on vehicles entering a city centre, with a complete ring of tolling stations located around the periphery of the charging area. The implications are more complicated because of difficulties associated with different journey purposes, emergency vehicles and residents, but the technology is available.

Road pricing has been advocated for many years by economic purists because it introduces the disciplines of the market to the road system, which is normally immune to such pressures. However, it is regarded as being extremely unpopular and no British government has been prepared to pass the legislation necessary for its general introduction.

Parking and access control

It has long been a requirement that any application for planning permission for development, whether commercial, industrial or domestic, should make clear the provision that is to be made for parking of motor vehicles and for access to the public highway system.

Thus, local planning authorities in considering any application will want to consider, in association with the highway authority, what provision has been made within the development. The general principle is that sufficient parking should be provided to meet all possible demand, and that junctions providing access should be designed in accordance with the relevant departmental standards.

Guidance for developers of housing estates is provided in the Department of Environment Design Bulletin No 32 – Residential roads and footpaths (DB32).[4]

Local planning authorities will publish a design guide, prepared in co-operation with the local highway authority, which will give details of the standards of parking provision required for residential development in their area. One of the most recent of such guides is the Surrey design guide[5] which aims to clarify certain issues and update them in the light of experience, but the main principles have not substantially changed. It is unfortunate, from the point of view of developers and designers, that different standards are applied in different local authority areas, but it does give the opportunity to each authority to take account of its own specific needs and opportunities. It would not be appropriate to apply identical standards in, say, the Metropolitan Borough of Calderdale, where new developments tend to be on either steeply sloping ground or land reclaimed from industrial contamination with those applied in, say, Cambridgeshire, where there has been little or no heavy industry and a gradient of 5% is considered to be a steep hill. Construction materials also vary widely across the country. Planners in Kirklees Metropolitan District frequently specify natural stone for all sorts of new development. The same requirement in Hertfordshire would simply price all development out of contention. Whilst building materials may not seem directly relevant to highway standards, engineers and planners should appreciate that the developer is concerned with the cost of the development as a whole, and the developer is able to trade off savings in, say, materials or density with costs in provision of parking. The developer is also concerned, as the planning authority should be, with the appearence of the development. Too many car parks have been built without regard to adequate landscaping and this does have an impact on the saleability, and hence price, of housing and on the viability of commercial developments.

There are certain common themes that run through standards of car parking required by local planning authorities.

The main objectives of the standards are:

1. To ensure the safety of people using the development from the danger of accident or attack.
2. To ensure the facilities are convenient to the prospective user.
3. To ensure that the future cost to the authority in maintaining the highway is kept within bounds.
4. To seek to contribute positively to the quality of life of residents and other users of the development.

In residential areas, the intention is to ensure that specific parking space is provided to meet the needs of residents for their own vehicles and additional space is provided for visitors and deliveries. Spaces are defined as being assigned or unassigned. Assigned spaces, i.e. those which are primarily intended for use by residents to park their own vehicles, will normally be within the curtilage of the property to which they are assigned and will always remain private. Considering each of the objectives (1) to (4) we can see that:

1. Safety is maximised: there is no opportunity for a child to run from behind a parked vehicle into the path of an oncoming car.
2. Facilities will usually be immediately beside the property
3. There is no future maintainance cost on the authority as the facilities are not adopted.
4. It is a matter of opinion as to whether a large estate of private houses each with its own driveway actually enhances the quality of life. The trend has tended to be away from this type of development, and towards rather more communal facilities.

Where the assigned spaces are contained within private property, the requirement for unassigned spaces will be greater. Surrey defines spaces as being 'near' or 'distant'. The precise definition is not very clear, although the situation described above would clearly be 'near', whereas a garage in a separate block along the road would clearly be 'distant'. Each dwelling with two or more bedrooms with one near parking space (a semi-detached house with single garage, perhaps) would require one additional unassigned space. If, however, the assigned space was 'distant', the additional requirement would be 1.25 unassigned spaces.

Many modern standards have been developed from the experience of the 1960s. At that time, emphasis was placed on the need to separate pedestrian and vehicular traffic, since it was perceived that the main danger to life was the conflict between vehicles and pedestrians. The main threat to quality of life was due to vehicle noise and emission of fumes. Lack of land was also considered a problem and high rise blocks were seen as a means of substantially increasing housing density whilst leaving sufficient space between the blocks for parking, play areas and grassing over. The resulting disaster has met with universal condemnation, but it is worth looking rather more closely at the reasons why it failed to meet the objectives defined above, particularly in the light of the rather different experiences of other European countries and the apparently successful refurbishments that are now under way in Sheffield and other cities.

Looking at each of our four objectives, we can see that one reason for the failure of the 1960s estates is related to parking.

1. Safety: it is a sad fact of life in Britain in the past two decades that crime levels are perceived to have risen dramatically. Whether this is due to the economic performance of the country or other social factors is outside the scope of this book. Police forces have been frequently drawing attention to the problems of car-related crime, the theft of valuables from vehicles or of the vehicles themselves, either for gain or 'joyriding'. Parking provision on a typical high

rise estate is distant vertically and frequently horizontally as well. It is often isolated, an ideal situation for the thief or vandal to do his worst.

Safety from physical assault, particularly for women, is a major concern. Isolated car parks, often badly lit, and the footpaths leading to them, are sometimes described as being a mugger's or rapist's paradise. Even if the reality of attacks is very rare, the fear is very real, and should be a taken into account by the designer of parking facilities.

2. Convenience: if a parking facility is not convenient for the user, then the user will not use it. Distant garages or parking lots, familiar on many high-rise estates, simply are not used. Residents and visitors park instead on the carriageway, causing obstructions, or on the verges, causing damage to kerbs and planting areas.

3. Maintenance: damage to kerbs identified under (2) has an immediate impact on the maintenance budget of the local authority, because they have to come and replace them. There is also the problem of vandalism. Again, the social causes of this phenomenon are beyond the scope of this book, but every engineer concerned with maintenance of 'bad' estates will be well aware of the costs involved in replacing signs, cleaning graffiti, keeping lighting operational and repairing other forms of malicious damage.

4. Quality of life: the subjective judgement of the quality of life of many who live on these estates is that it is somewhere round about zero. Certainly, it seems to have been beyond the wit of all governments to resolve the problems. Often, the basic problems with tower blocks are structural, sometimes worsened by asbestos contamination. They were badly built, inadequately controlled and financed without regard to the future. There were lots of lessons to be learned in matters such as the role of central and local government, and the professional role of engineers and planners. Hopefully, those lessons have been learned, and we will not make the same mistakes again. Some of the worst examples of structurally defective buildings – Hunslet Grange in Leeds, estates in Sheffield and Liverpool – have been cleared. But others, particularly in Sheffield, have proved themselves capable of refurbishment. The revitalisation of that city, led with some courage on the part of the City Council, shows just how important it is that the lessons of the past are learned and that we are willing to adapt our inheritance, however unsatisfactory, to modern needs.

One of the problems of achieving the four objectives outlined above is that it makes no reference to the effect on traffic levels on main roads. It is left to the developer to assess the housing market, and it may well be that maximum profitability is achieved by developing estates of four-bedroomed detached houses with double garages. The Surrey Design Guide would then require 0.5 unassigned spaces per property within the development, and the developer could cost everything out accordingly. There have been other profitable markets, often brought about by housing associations for, for instance, sheltered housing, which have very few traffic implications. No assigned parking space and one unassigned parking space, which could be the carriageway outside, may be adequate in this situation. However, the majority of speculative housing developments in the last ten years have been at the upper end of the housing market and could reasonably be expected to lead to ownership levels of 1+ cars per household. Many of these developments have been undertaken by small building companies in groups of 10–20 dwellings, and there is no mechanism for taking the cumulative effect of the development on the public road network into account when granting planning permission.

Box 22.1:

Case Study: A dilemma for the Council

A. Builder and Company applied for outline planning permission from Northtown Metropolitan District Council to construct ten, four bed-roomed detached houses in a field on the outskirts of the district. The field area was 0.7 ha, and it was effectively an infill plot in that Victorian housing was on one side of it, 1930s semi detached houses on a second, an old farmhouse and buildings which have recently been modernised, converted and sold on a third and a public highway, Southward Lane, on the fourth. The houses were to be in two groups of five, each round a shared surface cul de sac, giving two accesses on to Southward Lane.

The opposite side of Southward Lane had been developed in the 1960s, and a single footpath had been constructed at that time. Each house had a single garage and driveway to the front. The carriageway of Southward Lane was 4.5 m wide, bounded on the side of the proposed development by a stone wall 0.9 m high with a row of mature trees behind it.

The application included provision for widening Southward Lane on the side of the new development at the developer's expense to give a total carriageway width of 5.5 m along the front of the site, and to provide an additional footway 1.2 m wide. This involved taking down the existing wall and trees, and constructing a new wall on the highway boundary. A general site plan of the application is shown at Figure 22.1

Outline planning permission for residential use was granted but the planning authority reserved a decision on the nature of housing to be provided, expressing the view that small 'starter homes' would be more appropriate, given the needs of the area and the implications on the surrounding road network of more high car-ownership housing.

When the detailed application was submitted, it was still for ten four-bedroomed detached houses with double garages.

The occupants of the houses on the opposite side of Southward Lane objected to the application on the grounds that:

1. The proposals were inappropriate to the needs of the area.

302

Box 22.1 continued:

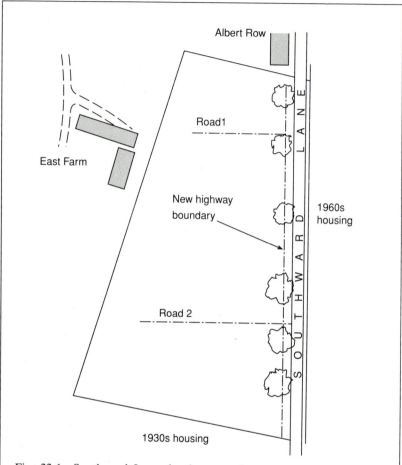

Fig. 22.1   Southward Lane development: layout

2.   The proposals involved removal of mature trees.
3.   The proposals would lead to unacceptable levels of traffic on the narrow roads leading to the site.

The objectors engaged an Engineer to advise them, particularly in respect of point (3).

The Engineer identified two particular issues to bring to the attention of the Planning Committee. These were:

1.   The visibility splays at the entrances to the two cul-de-sacs did not meet the standards laid down in the Authority's Design Guide. Widening Southward Lane on the development side aggravated the problem.
2.   The development would lead to a further marginal increase in traffic

Box 22.1 continued:

at a particular sub-standard junction approximately 0.5 km distant, which already had a significant accident record.

The Engineer was invited to meet the Planning Committee and explain the reasoning behind these ojections, which he did.

The Committee's decision was to reject the objections and approve the application, albeit with a rider requesting the developer to reconsider the decision to build large houses with double garages. The developer ignored this rider and built the houses as originally proposed.

This case study illustrates a number of points. Individual members of the Planning Committee informally told the Engineer that they agreed entirely with the points that he had made. However, they felt unable to turn the application down on those grounds because (a) they wanted more houses built, and even houses of the wrong sort would set up a chain reaction that would release cheaper housing in due course and (b) they were afraid that turning it down on general traffic grounds not directly adjacent to the site would lead to an appeal, which would be successful.

So planning authorities carrying out their functions of development control are not able to influence traffic levels directly through deciding which applications to approve. They are also not able to influence, other than indirectly, the traffic generation properties of new housing development.

The other reasons for the objection were over-ruled because (a) the highways department agreed to waive the visibility requirements in this particular case and (b) the developer agreed to plant trees to replace the ones that were being taken down.

The amount of parking to be provided at commercial and industrial development is also determined at planning application stage. It is normally worked out on the basis of one space per so many square metres of usable floor space within the development, depending on the use to which the building is to be put. Again, individual authorities will have their own standards, which will generally vary according to car ownership levels in the area.

Eastman and MacKenzie[6] of JMP Consultants have undertaken research into requirements for parking for various commercial land uses, based on the TRICS database which JMP manage on behalf of a consortium of local authorities, the purpose being to base design of facilities on actual practice. Their study of the database has given the results shown in Table 22.1.

Table 22.1   Parking demand for different land uses (Source: C Eastman, JMP Consultants Ltd)

Sq. m of Gross Floor Area per parking space	Average	85%ile
Food superstores	16	12
DIY superstores	27	19
Retail parks	33	33
Offices	42	23
Business parks	59	33

Based on this information Eastman and MacKenzie have drawn up a range of recommended values on the basis that if a standard of average demand + 1 standard deviation is taken as the requirement, the car park would overspill on 17 occasions in a year. A better standard might be to provide for average demand + 2 standard deviations to obtain the figures in Table 22.2.

Local authorities face a substantial dilemma when it comes to determining parking provision. For many years, it was deemed that, in small centres, the provision of ample free parking was essential to the economic well-being of the centre. In major centres, limiting parking availability would act as a rough and ready control on the amount of traffic using radial routes to reach the centre and therefore obviate the need for increasing capacity on those routes. Parking would be provided around the fringe of the city centre and differential charging used to discourage drivers from coming too far in. Planning policy would be used to discourage the provision of additional private city centre parking and pricing mechanisms used to influence short stays who might be expected to be shopping or on business.

This was of limited effect in seeking to overcome the problems of city centre congestion. It was found that the number of parking spaces within a central area which were not controlled by the local authority was usually substantial, partly because at an earlier time planning consents had insisted on developers providing off street parking, often in basements. It has also been found that companies are prepared to pay out almost unlimited amounts in parking charges in return for having their employees able to park almost outside the door. Although the taxation position of company cars has varied considerably in recent years, it has become very apparent that other methods were necessary to control congestion. It is no longer sufficient to simply rely on parking charges and planning consents.

The policy of restricting parking in town centres has helped to accelerate the move towards out of town shopping centres and business parks. Many would argue that this is undesirable, particularly where it places pressure on Green Belt land. It can be seen to be contrary to government and local authority policies to revitalise inner city areas. The evidence, however, is not always conclusive. Gateshead Metro Centre has not led to an apparent decline in shopping activity in

Table 22.2 Suggested parking standards (Source: Eastman)

Recommended sq. m of Gross Floor Area per parking space

Food superstores	9
DIY superstores	15
Retail parks	25
Offices	20
Business parks	30

Newcastle City Centre, although Meadowhall has caused some problems in Sheffield.

Box 22.2:

Case Study: The parking problem

Winterfield is a small town, population 6500, on the fringe of a major urban area. The local highway authority is Westborough Metropolitan Council, which is based in Westborough Town Centre, about eight miles from Winterfield. There is substantial commuter traffic from Winterfield to Westborough.

Parking in Winterfield has always been free. A lot of it is on street and in the Market Square, amounting to 150 spaces in total, and there is room for approximately 250 cars parked on a vacant, unsurfaced site close to the Market Square. On-street parking is subject to a two hour limit, and is patrolled on a random, occasional basis by a single warden, who arrives by bus from other parts of the metropolitan district. Minimal waiting restrictions are in force, and it is not unusual for congestion to be caused by inappropriately parked vehicles. There are residential properties in the area immediately outside the town centre, and residents frequently complain that they are unable to get access to the front of their property because of parked cars.

Most of the properties in the area of the Market Square are either retail shops or commercial premises. A planning application for a supermarket adjacent to the waste ground car park has recently been turned down on the grounds of inadequate parking and access. The same company, however, is constructing a major superstore about two miles from Winterfield.

Westborough Council was found, in the last financial year, to have had a deficit on its parking operations of £267,000, and it is therefore obliged to take action in order to meet its legal obligations. The appropriate committee had three alternative courses of action available to it:

1. To seek competitive tenders from alternative contractors for the management of its car parks.
2. To raise car parking charges on those parks where they were already levied by the minimum amount possible to make good the deficit, without incurring any additional expenditure. This would have involved a general rise of 15% rounded up to the nearest 10p.
3. To raise charges generally by 25% and impose them on those parks where there were no charges, and to use the additional revenue to commence implementation of a rolling programme of improvements to surfacing, lighting, security and environmental conditions. It was not anticipated that increases of this size would lead to any deterrent effect and that parking demand would remain constant.

The Council resolved to adopt Option (3). As far as Winterfield was concerned, this implied charges of 30p per hour with a maximum stay of 2 hours for on street parking and 10p per hour unlimited for off street parking. Charges were to be collected using pay and display machines

Box 22.2 continued:

and patrols were to be made more frequent. In the short term, no improvements were envisaged other than those implicit in the closer control of on-street parking. In the medium term (3–5 years) the unsurfaced, off-street car park would be surfaced, additional lighting would be provided and a planting scheme implemented.

A substantial amount of protest erupted when the Council's plans were announced. The complaints could broadly be listed as follows:

1. Any parking charges would lead to a loss of trade, particularly once the new superstore opened.
2. There was no provision, other than by payment of about £5 per week, for people who worked in the town to park all day whilst they were at work. Comparison was made with the 'commuter' car parks on the fringe of Westborough Town Centre, where charges of 40p/ day were levied.
3. Residents in streets close to the centre of Winterfield would find their position made even more difficult as parkers sought to obtain free parking in residential streets.
4. Local people were having charges levied for no apparent improvement in service.

The Council implemented the proposals as planned but agreed to review the position after three months. At this review, it was found that:

1. There had been a substantial drop in parking both on-street and off-street, and this was leading to a revenue shortfall for the authority. There had not, however, been a noticeable drop in trade.
2. Additional congestion had been caused on streets leading into the Market Place due to motorists taking advantage of free on-street parking a comparatively short distance from their destination. This had caused difficulties for residents. There were signs, however, that this effect was wearing off as people got used to the idea of paying.

Faced with this position, the authority, after discussions with local business and community leaders, agreed:

1. To reduce the charge for parking in the Market Place, but not elsewhere on-street, to 10p/30 minutes, with a maximum stay of 1 hour.
2. To extend the zone covered by on-street parking regulations and introduce, where appropriate, a residents only scheme.
3. To bring forward proposals to install lighting on the off-street car park for implementation in the next financial year.

The local authority can generally be seen to have done a reasonably good job. Apart from the legal requirement for it to break even on its car parking accounts, there is no particular reason why car owners in general should be subsidised out of public funds. The cost of a parking space varies according to the value of the land on which it is located, but with city centre land fetching rentals of the order of £100/m^2 and a car space with associated aisles and accesses taking up around 22 m^2, the cost of the land alone can be reckoned as about £2000 per year. If a

Box 22.2 continued:

typical city centre car park is in use for 200 days/year, parking costs need to be around £10 per day just to pay for the land. Outside major urban areas, of course, land costs are much cheaper, but the £5 required of commuters in Winterfield can be seen as barely meeting the costs they impose, and the 40p per day in Westborough Town Centre is very reasonable.

There is a balance to be drawn when faced with competition between an existing centre with limited and expensive parking and a modern retail park, where parking is usually provided at no direct cost to the user – although, of course, the user still pays through higher charges for the goods being bought. Basically, the high turnover and greater operational efficiency of modern stores allows them to provide free parking (usually on low cost land) without becoming uncompetitive. Whether the reduced choice that is involved in using out-of-town shopping malls will eventually restrict their growth remains to be seen.

The Council, however, did not get it all right. They should have foreseen the need for very short stays in the Market Place and fixed the pricing accordingly. Although it is a simple matter to change the settings of a pay-and-display ticket machine, and to repaint the notice boards, 'getting it right eventually' is not a good approach to engineering design. It tends to give the designer a bad name and to lead to uncertainty among motorists.

The Council should also have been more sympathetic to the needs of residents. A common problem for people living in urban centres is that of cars parked on the highway outside their homes. At best, this is unsightly and inconvenient and at worst is positively dangerous. By drawing the cordon of the controlled zone too tightly round the central area, the Council inevitably caused problems for those streets immediately outside the cordon, which happened to be residential. Residents parking systems are not ideal, in that they do not make very good use of available road space, and they are difficult to enforce, but they do go some way to address the problem. A study of attitudes to this particular problem has recently been commissioned by the Transport and Road Research Laboratory.

The fact that trade levels did not fall off should not have come as a surprise. Imposition of parking charges are frequently seen as a threat by shopkeepers, and this may seem logical given the priority that major supermarket chains give to providing free parking. However, the cost of parking is only one factor, and probably quite a small one, in a decision to visit a particular facility. Range of goods, price of goods, convenience and environmental factors are all important. Small town centres score well on factors like personal service, fresh food, variety and (often) price: paying a few pence for parking is unlikely to put anyone off. It may be that other matters, like machines only taking one type of coin and not giving change, or paystations remote from parking stalls, are more likely to discourage motorists than an extra 20p.

A start has been made in the United States on the management of demand by means of a 'Commuter Plan', prepared by city centre employers. This requires that an employer works towards a situation in which his workforce brings a reduced number of cars to work. He might start with a ratio of 1.25 jobs/car and be required over, say, 5 years to move to a ratio of 1.75 jobs/car. He can effect the change by a number of means: changing work locations or hours, putting on a company bus or subsidising public transport tickets. It is difficult to see such a system being accepted in Britain, although something like it could be taken into account when granting planning consent for new developments.

Park and ride

Park and ride is not new. In the 1920s and 1930s, developers were quick to take advantage of the opportunity offered by the extensive rail network, particularly in London but also in other major cities. Their customers were offered the dual facility of living in the countryside while being able to travel easily to the city for work, shopping and recreation. With the rapid growth in car ownership through the 1960s and 1970s, the provision of a car park – often on land made redundant by the withdrawal of freight, coal and parcel handling facilities – became an essential part of a railway station.

There have been three separate markets identified as having potential for park and ride. The initial impetus came from the commuter. Wage rates in city centres, particularly for office and shopworkers, were not adequate to allow them to buy their own property close to their workplace. Land prices, in any case, were such that only the very richest members of society could hope to own even a small flat in central London. So commuting became an everyday part of life in London and other major cities.

With the improvement of rail technology and the extension of the network, and consequent reduction in journey times, the commuter zone around London steadily increased. Electrification of the Southern Railway's Brighton Line brought the South Coast within reach. Extensions to the 'underground' (usually on the surface) into Essex and Hertfordshire were specifically aimed at commuters. More recently, the introduction of Intercity 125s on the Great Western brought Bristol within commuter distance of London and the electrification of the East Coast Main Line has brought demand for season tickets from as far away as Doncaster, Wakefield and York. It is generally thought that the opening of the channel tunnel and the associated high speed link to London will bring the first international commuters from the Pas de Calais region of northern France.

Unfortunately, commuter traffic is not particularly profitable for transport operators. The peaked nature of the demand means that a large amount of capacity, in the form of track space, trains, stations and signalling, has to be provided and can only be used for two or three hours on five days per week. This is true of any transport system, and it has been apparent for at least the last twenty years that it simply is not economic to try to cater for peak flows on the road system. On roads, however, where most vehicle movement is only subject to minimal controls, individual drivers have a lot of freedom to determine how they will cope with congestion, and if the worst comes to the worst, they simply have to wait. A railway, however, has to run to a timetable, and it is necessary to make some form of provision for peak traffic if public safety is not to be unacceptably compromised.

Thus, from quite early times, the railway companies attempted to attract different customers, who could use the spare capacity on the commuter networks when the commuters were not using them. This primarily worked through ticketing systems, whereby off-peak travel could be undertaken at a substantial discount. Cross ticketing between different operators is important if these schemes are to be successful and in large areas such as London zoning systems are appropriate. To some extent, deregulation of bus services has made the application of off-peak travelcard type tickets more difficult, but a clear market exists for shopping and leisure travellers to use public transport, including park and ride facilities.

The third potential market for park and ride is the business traveller, particularly en route for London. Most provincial centres in England and Wales now have an Executive style rail ticket on morning trains to the capital, and these frequently include use of reserved car parking spaces at the departure station. In Scotland, similar facilities exist but the airlines tend to be able to gain an advantage over rail because of the journey times involved. The range of 100–300 miles is appropriate for this type of travel by train, while over 300 miles (or journeys involving a sea crossing) will tend to be by air.

British Rail has exploited the market for park and ride in reverse, particularly at centres outside London. Businessmen will often require the convenience of a car at the end of their journey, so franchise arrangements have been set up with a major car hire company whereby a car can be collected at the destination station.

So park and ride began as a means of widening the range of opportunity for commuters, was extended through ticketing systems that sought to make efficient use of spare capacity and was adopted from the mid-1960s onwards as a means of encouraging business travellers to undertake the longest leg of their journey by rail or air. There were almost no successful examples of a non-rail based system because the additional speed on the rail leg was necessary to make up for the inconvenience and time loss involved in changing modes. Early attempts to introduce bus-based park and ride systems, such as Nottingham or Southampton in the late 1970s, could be said to have failed because the buses were still caught in city centre congestion and there was no time or cost advantage to motorists to leave their car on the outskirts. Buses suffer from an image, often unjustified, of being dirty and inconvenient and it is only recently that some operators have managed to overcome this problem.

As car ownership has increased the degree of congestion in city centres has also increased. With major road construction in urban centres ruled out by economic and political considerations, planners have again turned to park and ride as a potential solution, possibly the only solution. It has to be seen in the context of transport planning for the city as a whole, and may involve extensions of bus priority measures to ensure that journey times by bus are at least comparable with those by car. Although commuting, with its waste of valuable city centre land for parking, is one target for park and ride, other journeys, such as shopping and leisure, are also included. There has been a tendency to implement park and ride schemes in historic cities, where redevelopment of the city centre to allow for new roads or car parks, is out of the question. Bath, Oxford, York and Chester are all implementing extensive schemes.

The Government's view is that park and ride can make a positive contribution to alleviating congestion and improving the environment. Roger Freeman MP, Minister for Public Transport, speaking at the English Historic Towns Forum in Oxford on 18 September 1991 said:

> In the right circumstances, park and ride can make a valuable contribution to the alleviation of congestion and to the improvement of the environment, both of which are amongst the Government's policy objectives.

From the beginning of the 1980s bus and rapid transit based park and ride has been seen as having major potential benefits for urban areas. Most of them started as seasonal schemes to cope with either the pre-Christmas shopping rush or to cope with summer holiday traffic in congested locations such as Fowey, Cornwall. The cost of the bus journey was either met entirely by the local authority promoter of the scheme or was included in the price of the car park ticket.

Around 36 park and ride schemes were introduced in Britain during the period 1981–90, of which 29 came in the period 1987–90; 49% were seasonal, 33% during the pre-Christmas period and 16% in the summer.[7]

The principle of park and ride revolves around the establishment of a cordon around the city centre with well designed car parks located at points where main radial routes cross the cordon. A frequent, reliable bus (or LRT) service connects the car park with convenient locations in the central area. The cost to the user must be demonstrably less than the cost of a central car park, and the service must be seen as being more convenient and, preferably, quicker. High quality must be associated with all aspects of the service, so modern, clean, comfortable vehicles are necessary but so are polite staff, pleasant waiting areas and apparent security.

The optimum location of the cordon will vary from city to city. In

(a)

(b)

Fig. 22.2 (a) & (b) York Park and Ride (Courtesy York College of Further and Higher Education)

the case of Bath[8,9] it was found to be the administrative boundary of
the local authority area. This was because this approximately coincided
with the extent of the built-up area and the commencement of the
Green Belt, and because of potential conflict with adjacent authorities
as regards planning consents. In other areas, a ring road or similar
facility might provide an appropriate cordon. Extensive development
on the fringe of a city, in which adjoining communities have been
absorbed, perhaps along particular corridors, will mean that care has
to be taken to determine the position of the cordon.

A plan showing the location of Bath's park and ride sites, existing
and proposed, is shown at Figure 22.3.

The optimum capacity of the system is not easy to determine.
Insofar as figures are readily available, usage of park and ride schemes
is steadily increasing (see Figure 22.4) but there is almost certainly an
element of only recording success. What is clear is that the potential
for greater use of park and ride is there and this can only increase as
congestion in city centres increases.

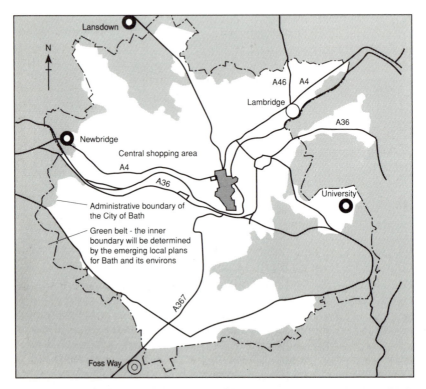

Fig. 22.3 Bath Park and Ride area (Courtesy Thomas Telford Ltd/R D
MacPherson by kind permission of the Director of Property and Engineering
Services, Bath City Council)

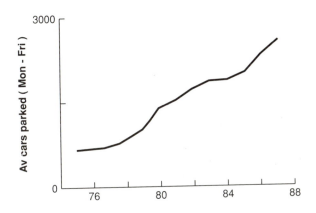

Fig. 22.4 Usage of Oxford Park and Ride (Source Department of Transport Traffic Topics 5)

Careful design of the parking facility is essential if use is going to be maximised. Suitable surfacing, which might include hot rolled asphalt for heavily trafficked access roads and some combination of concrete block paving and surface dressing for aisles and parking stalls, is essential. Good positive drainage, totally eliminating the possibility of standing water, is required. Attractive landscaping should be incorporated into the layout of the car park, but this must be consistent with security requirements, which mean that the whole park must be visible from a small number of vantage points. Continual manning during the period when the park and ride system is operational, is seen as being necessary if problems of theft and vandalism are to be kept under control.

The whole of the site should be well lit and there should be a substantial shelter, a public telephone and an information board. More extensive facilities are now being installed on new park and ride systems.

Figure 22.5 shows the layout of the proposed park and ride facility at Foss Way, Bath. Note the following points:

1. Parking areas are divided up by planting into relatively small sections.
2. Facilities provided include toilets and information point.
3. Disabled parking spaces provided. This implies consistency in access to vehicles.
4. Standard roundabout provided at access from the main road, in accordance with Departmental Advice Notes TA23/81 and TA42/84.
5. Extensive planting around the perimeter of the site at a cost of additional parking spaces.

314

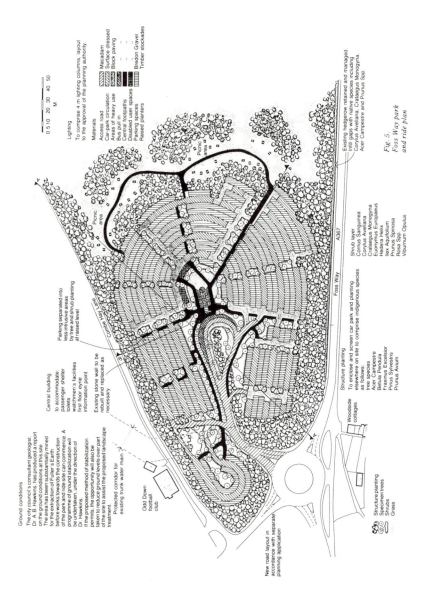

Ground conditions

The city council's consultant geologist,
Dr A B Hawkins, has produced a report
on the ground conditions at this site.
The area has been substantially mined
for the extraction of Fuller's Earth
before works towards the construction
of the park and ride site can commence. A
programme of ground stabilization will
be undertaken, under the direction of
Dr Hawkins.
If the proposed method of stabilization
permits, the opportunity will also be
taken to reduce ground levels over part
of the site to assist the proposed landscape
treatment.

Protected corridor for
existing trunk water main

New road layout in
accordance with separate
planning application

Structure planting
Specimen trees
Shrubs
Grass

Odd Down
football
club

Central building
to accommodate:
passenger shelter
toilets
watchmen's facilities
first floor eyrie
information point

Parking separated into
less intrusive areas
by tree and shrub planting
at raised level

Existing stone wall to be
rebuilt and replaced as
necessary

Structure planting
To enclose and screen car park and planting
elsewhere on site to comprise indigenous species
as follows:
tree species
Acer Campestre
Betula Pendula
Fraxinus Excelsior
Pinus Sylvestris
Prunus Avium

shrub layer
Cornus Sanguinea
Corylus Avellana
Crataegus Monogyna
Euonymus Europaeus
Hedera Helix
Ilex Aquifolium
Prunus Spinosa
Rosa Spp
Viburnum Opulus

Woodside
cottages

Foss Way

A367

Combe Hay Lane

Picnic
area

Picnic
area

Lighting
To comprise 4 m lighting columns, layout
to the approval of the planning authority.

Materials
Access road
Car-park circulation
Areas of heavy use
Bus pull in
Central footpaths
Disabled user spaces
Parking spaces
Raised planters

Macadam
Surface dressed
Block paving

Bredon Gravel
Timber stockades

0 5 10 20 30 40 50
M

Existing hedgerow retained and managed.
Infill gaps with native species including
Corylus Avellana, Crataegus Monogyna,
Acer Campestre and Prunus Spp.

Fig. 5.
Foss Way park
and ride plan

Fig. 22.5 Foss Way Park and Ride, Bath (Courtesy Thomas Telford Ltd/R D MacPherson by kind permission of the Director of Property and Engineering Services, Bath City Council)

6. Separate defined footpath system between parking areas and central area.
7. Bus path through entry, stand and exit is continuous and always forward.
8. Comments on ground stabilisation and resulting opportunities for adjusting the ground profile.

At York, the Dringhouses Park and Ride facility makes use of an out of town superstore location for a shared car park. The park and ride facilities are on the opposite side of the car park from the superstore, so different spaces could be said to be allocated to each use. Access is from a roundabout on the main A64 Leeds and Tadcaster road a short distance inside the cordon formed by the York Outer By-Pass. Facilities provided have been extended to include toilets and a tourist information centre, and a petrol station is provided as part of the superstore development. The College of Arts and Technology is also located in Dringhouses, providing additional potential customers for the bus service.

The location of the Dringhouses scheme is highly visible as it is on redundant railway land immediately adjacent to the main railway into York from the south and beside a main radial road into the city.

Park and ride can sometimes only meet its full potential in the context of overall transport planning for the future. Proposals for Chester sees the development of the scheme there in the context of new road construction, possible LRT and future bus lanes.

Norwich is another city where park and ride is seen as having potential for solving urban traffic problems.[10] (See Figure 22.6). An additional feature in the Norwich strategy is the Residential Area Protection Scheme, which is descended from the earlier ideas advanced in the original Buchanan Report, Traffic in Towns, which used Norwich as one of its examples.[11] The report envisaged dividing the city centre into a series of 'rooms', with 'barriers' between them. Pedestrian and cycle traffic, together with some buses, would be able to cross the barriers, but other traffic would only be able to gain access direct from the primary road network.

Park and ride would appear to offer potential benefits in resolving some of the problems of urban congestion, but there is, as yet, insufficient evidence to establish how general its application might be.

References and further reading – Chapter 22

1. Tyson W J. *Planning and financing Manchester Metrolink*. Journal of the Institution of Civil Engineers: Transport. 95. Aug: 141–150. August 1992.

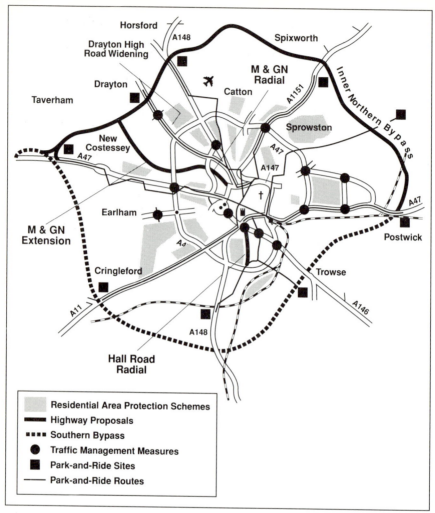

Fig. 22.6   Norwich Traffic Strategy (Courtesy Institution of Highways & Transportation)

2. Bodell G and Huddart K W. *Tram priority in Hong Kong's first Light Rail Transit system*. Traffic Engineering and Control. 28. 9: 446–451. September 1987.

3. Chin K, Mundy J and Thompson T. *Control of the LRT/traffic conflict: an update from Hong Kong and simulation using the FLEXSYT program*. Traffic Engineering and Control. 33. 2: 65–71. February 1992.

4. Department of the Environment/Department of Transport. *Design Bulletin 32 – Residential roads and footpaths – Layout considerations*. HMSO. 1977.

5. Surrey County Council. *Roads and Footpaths – A design guide for Surrey*. Designers' Handbook. 1990.
6. Eastman C. *A question of standards*. Parking Review. Nov/Dec 1991.
7. Statistics drawn from Department of Transport. *Traffic Topics 5. Park and ride*. London. 1991.
8. MacPherson R D. *Park and ride: progress and problems*. Proceedings of the Institution of Civil Engineers: Municipal Engineer. 93. 1: 1–8. Paper 9905. March 1992.
9. MacPherson R D. *Implementing park and ride*. Paper presented to Construction Industry Conference Centre conference 'Passenger Transport and Urban Survival'. Oxford. 1991.
10. Ramsden J, Coombe D and Bamford J. *Transport Strategy in Norwich*. Highways and Transportation. 39. 2: 3–9. February 1992.
11. Buchanan C (Chairman). *Traffic in Towns*. Report of the Steering Group. HMSO. London. 1963.

Chapter 23

Europe: The DRIVE Research Project

DRIVE I

In June 1988, the European Commission adopted a major research initiative into the use of telematics in order to assist road traffic management. This was the DRIVE project, the first phase of which was completed in 1991.

In order to continue the work, a further phase, DRIVE II, was set up to run through 1992–94.

DRIVE resulted from a series of research studies undertaken on behalf of the European Commission from 1985 onwards to see whether the adoption of informatics could make a significant contribution to improving safety in road transport. These studies confirmed the Commission's view that there was an urgent need for a programme of work in this area.[1]

The objectives of the DRIVE programme as originally adopted were to improve road safety, transport efficiency and environmental quality. These follow from the definition of the negative effects of road transport which can be summarised as:

- 55,000 people killed, 1.7 million injured and 150,000 permanently handicapped in road accidents in the European Community countries each year.
- a cost of accidents estimated at more than 50 billion ECUs (£36 billion) every year plus an untold amount of human misery and suffering.

- a cost estimated at 500 billion ECUs (£357 billion) each year due to congestion and other traffic costs.
- a contribution to environmental pollution estimated to cost Europe between 5 and 10 billion ECU per year.

DRIVE I projects were grouped under four main headings:

1. *General approach and modelling*
   In order to carry out meaningful evaluation of systems and policies, sound quantitative estimates of their cost and performance are required as well as some way of testing their effects in terms of transport efficiency, safety and pollution. Social costs and benefits must be assessed against common criteria.
   Fifteen projects in this group were concerned with establishing procedures for modelling the effects of Road Transport Informatics (RTI) on demand, simulation, environmental effects, evaluation and implementation.

2. *Behavioural aspects and traffic safety*
   The improvement of road safety was the original objective that led to the establishment of DRIVE. Projects in this group were concerned to look at hazard and accident data analysis, the safety of vulnerable road users (pedestrians and cyclists), the evaluation of behavioural analysis related to RTI implementation, collision avoidance, data recording, and the wider impact of RTI on travel patterns.

3. *Traffic control*
   Traffic control refers to the process of maintaining a smooth flow of traffic by measures such as altering the timing of traffic lights in response to information available to controllers. The aim of the DRIVE project is to develop an Integrated Road Transport Environment (IRTE) in order to increase capacity of road networks and to generally improve efficiency.
   IRTE involves the collection of data, transmission to central control, analysis, adoption of appropriate strategies and dissemination of resulting information to drivers.

4. *Services, telecommunications and databases*
   A diverse set of projects are included under this heading. These are grouped in five clusters: public transport; freight management; digital maps and databases; information and broadcasting systems; and communications. They are basically concerned with the storage and transmission of information of particular relevance to fleet operators.

DRIVE II

The output of the first phase of DRIVE is 73 project reports covering the range of topics outlined above.

The central function of DRIVE II is

to contribute to the development in the field of transport of integrated trans-European services using advanced information technology and communication to improve the performance (safety and efficiency) of passenger and goods services and at the same time reduce the impact of transport on the environment.[2]

More than 500 organisations are involved with DRIVE II including government, administration, industry, universities, research establishments, service providers and users. It is anticipated that it will take up 20,000 person-months over 3 years. It is divided into 56 separate projects.

Most of the projects are located in seven separate operational areas, although there is some work which crosses boundaries. Full details are given in the appropriate EC publication.[3] The seven areas are:

1. Demand management
    There is a need for planning authorities and transport managers to strike a balance between the demands of travellers, the capacity of the system and the resources available to expand the system. The main goal is to enable travellers and fleet operators to make the choice best satisfying their needs, considering the duration of the journey, the cost to them and the time at which the journey takes place.
    The area includes control of the use of road space, access and parking; direct charging for the use of the road; tolling; policing and enforcement; links with public transport through common ticketing.
2. Travel and traffic information
    This area comprises the collection, processing and distribution of road traffic information to home, office or mobile receiver. This will be relevant in several other areas which are dependent upon the transmission of information.
    The main goal is to provide drivers, fleet managers, public transport operators and users and pedestrians with attractive, user-friendly, affordable and reliable information. This information needs to include:
    Traffic information on current traffic flow, availability and cost of parking and forecast traffic flow due to roadworks and other temporary situations.
    Information relevant to trip planning, which might include alternative modes available, weather conditions, tourist information and likely hold-ups due to roadworks or sporting events.
    Information related to position fixing and route guidance.
3. Integrated urban traffic management
    This area is concerned with traffic management and control over the whole network, and can therefore be seen as a development of

existing Urban Traffic Control systems. It involves collection, transmission, analysis and dissemination of data in real time so as to be relevant to users. A central control can optimise choices on route guidance, parking management, emergency management, environmental management and tidal flow systems. Dissemination of information can be through signal systems, variable message signs and other roadside equipment and in-vehicle equipment.

4. Integrated inter-urban traffic management
    This area covers traffic control and driver information systems on motorways and other high speed roads. Incident detection is important, possibly involving automatic systems based on video image processing. One-way and two way communication with the driver are necessary, with developments of the type of variable message signs now becoming common on motorways.

5. Driver assistance and co-operative driving systems
    A common component of every road accident is the involvement of a driver. It therefore follows that, if the driver's task is made easier, there should be a reduction in the level of road accidents, which is one of the central objectives of the DRIVE programme. The main goals of this area are therefore to improve traffic safety, improve road capacity, adapt information to driver needs, assist the driver's task and improve comfort and address topics relevant to vehicle control which could be subject to standardisation.

6. Freight and fleet management
    The goal is to demonstrate the operational and commercial benefits of the use of advanced information and telecommunication techniques for improving the planning and management of freight operations throughout Europe. The logistics of freight operations can be extremely complex and involve several participants. There will be a consignor and consignee at the beginning and end of the transport chain, but in between there may be forwarders, carriers and operators. There will also be a parallel system of administration relating to despatch, delivery and charging.
    Efficiency is dependent on fleet management, including route planning and vehicle and driver scheduling and on vehicle management including condition monitoring and maintenance, tachographs and other on-board monitoring. Better use could be made of rail and ferry facilities, avoiding the long delays which are currently experienced, by the adoption of a multi-modal booking system.

7. Public transport
    The only apparent answer to congestion in urban areas and in main inter-urban corridors appears to be greater use of public transport, particularly for journeys to work. This in turn implies an attractive public transport system that offers sufficient capacity, modern levels of comfort, reliability, ease of use and reduced travel times.

Applications in DRIVE II include better methods of operational planning and scheduling based on more reliable databases of user behaviour and requirements, better methods of user information and better methods of fare collection.

Within these seven areas, the project aims to address three main groups of activities:

1. The definition of functional specifications for the use of technology and telematic systems for communication and traffic control.
2. The development of new technologies and experimental systems.
3. Validation of the output through a series of pilot projects in cities and corridors throughout Europe, covering both passengers and freight.

The largest part of the resources devoted to the project is concerned with part (3).

Participating organisations from all twelve countries in the European Community, together with Austria, Norway, Sweden and Finland are involved in DRIVE.

References and further reading — Chapter 23

1. Commission of the European Communities DG XIII – *Telecommunications, Information Industries and Innovation*. The DRIVE programme in 1991. Brussels. April 1991.
2. Keen K. *EC Research and technology development in Advanced Road Transport Telematics, 1992–94*. Traffic Engineering and Control. 33. 4. April 1992.
3. Commission of the European Communities DG XIII – *Telecommunications, Information Industries and Innovation*. Research and Technology Development in Advanced Road Transport Telematics in 1992. Brussels. April 1992.

Chapter 24

Traffic Calming

Background

In recent years, one of the most significant developments in traffic engineering in the United Kingdom has been the advent of traffic calming.

The intention is to reduce the speed of vehicles in locations where they are likely to come into conflict with pedestrians and cyclists. In 1990, a total of 5217 people were killed and 60,441 seriously injured on roads in Great Britain. Of these 1676 fatalities and 17,142 serious injuries were to pedestrians. A pedestrian has far less protection than a car occupant and, in a collision, is therefore far more likely to be injured.

The Government's estimate of the cost of each road accident in 1990 was:

Fatal	£682,467
Serious injury	£24,717
Slight injury	£2,266

The total cost of road accidents in Great Britain was £5.8 billion. Clearly, if the number of pedestrians killed or seriously injured in road accidents can be significantly reduced, this will lead to a dramatic drop in costs, as well as relief from the grief and unhappiness that is the inevitable result of a road accident.

Ninety-five per cent of accidents to pedestrians occur on urban roads having a speed limit of 40 mph or less; 35% of these involve children aged 14 or under. The seriousness of likely injury to pedestrians varies according to the speed of the vehicle. If the vehicle is travelling at less than 20 mph, most pedestrians will survive. At 30 mph, 50% of pedestrians will be killed; at 40 mph most pedestrians will be killed.

Very similar statistics can be produced for pedal cyclists who, like pedestrians, have very little protection in the event of a collision with a vehicle. Although the number of cyclists is small compared with the number of pedestrians, it is rising and there is reason to suppose that increasing concern for problems of pollution will lead to cycling becoming a more significant mode of transport in the future.

It is the policy of the Government to bring about a reduction in the number of road accidents by one-third of the 1990 level by the year 2000 and this policy has implications for the amount of support offered to local highway authorities through Transport Supplementary Grant.

The statistics show quite clearly that if the maximum speed of vehicles in urban areas can be reduced to below 20 mph, there will be a dramatic reduction in both the number and seriousness of accidents in general and those involving pedestrians in particular. There will also be an improvement in environmental quality brought about by reduced noise and vehicle intrusion. The main objective of traffic calming, as seen by local authorities and the Department of Transport is to bring about a reduction in the maximum speed of vehicles in urban areas to 20 mph. This does not necessarily mean that there should be a reduction in the speed of all vehicles, only in that of the fastest. Measures should be taken to ensure that all vehicles travel at or below the design speed, which might be 20 mph on a local distributor which also functions as a shopping and residential street, and 6–10 mph on residential cul-de-sacs and shared surface courts.

Until recently the basis of urban planning has been to design around the needs of the car, leaving pedestrians to either stay at home, go the long way round, run the risks of crossing a road provided for high speed traffic or of being attacked in a dark and smelly underpass. There was an implicit assumption that car ownership could just go on growing, so we would have to rebuild our cities accordingly. Even if this theory has seemed, for twenty years now, to be becoming increasingly untenable, it is only very recently that we, in Britain, have begun to think of alternatives to cities dominated by ever wider roads.

An excellent summary of the development of thinking in traffic planning is provided by Rodney Tolley in his book, *Calming Traffic in Residential Areas*.[1] He quotes a London policeman, Alker Tripp, pointing the way forward:

> It is wrong to have local groups of population clustering about conduits that carry high speed traffic . . . the only answer is replanning.[2]

Buchanan, in what is largely regarded as the report that started the science of traffic planning,[3] recognised the problems, indeed, the impossibility of redrawing our cities to accommodate even a proportion of traffic demand. But he still produced a report filled with glossy coloured plans of new urban expressways carving through old residential areas. He did, however, point to the need for car restraint, for recognising the different functions of different roads, and for enhancing environmental quality.

After Buchanan, nothing very much happened that could be described as enlightened treatment of traffic problems for at least twenty years. Town centres continued to be flattened to create boring shopping precincts with nearby car parks. The number of Inner Relief Roads under construction gradually reduced as their costs escalated and the number of objectors to them grew. Congestion grew worse, but motorists proved themselves to be remarkably tolerant of delays and willing to pay penal parking charges in order to be able to drive their car to the office every day. Major retailers realised that they could build huge superstores on the outskirts of cities and provide car parks nearby. Brent Cross, Gateshead Metro Centre and Sheffield Meadowhall were born, but it must mean a further decline in the prosperity of the Inner City and a diminution of the facilities available to that substantial part of the population that does not have access to a car. It can only tend to aggravate problems in residential areas.

As Tolley[4] puts it:

> The comparison (with the Netherlands and Germany) will not show British progress in a very flattering light, for although *Traffic in Towns* was hugely influential in its day, nothing of significance has been added to it in the UK in a quarter of a century. . . . the centre of gravity of ideas, research and innovation now resides firmly south and east of the North Sea.

The planning of new residential areas has been dominated by the need to separate vehicular traffic from pedestrians in what has become known as a 'Radburn' layout, after the location of its original application in Radburn, New Jersey, in 1928. Many 1960s developments follow this pattern, with distributor roads serving cul-de-sacs and courts around which the houses are grouped. There is frequently a separate system of footpaths and cycleways, as in Milton Keynes or Stevenage, and it should be possible to get from any housing area to any facility without crossing a road, other than on a bridge. In many ways, Milton Keynes is a successful application of this system, but it does spread over a very large area. The product may not be so pleasing when it is used to give access to high density developments and tower blocks.

The Department of the Environment Design Guide to Estate Layout[5] (DB32) follows on from the Radburn idea of a hierarchical structure of estate roads and a separate system of footpaths and cycleways but it moves away from the idea of separate parking areas of windswept concrete. Parking provision is seen as crucial, and the 1992 edition of DB32 points to the fact that planned parking provision will be largely ignored if it does not acknowledge the public preference for being able to park close to the front door in a highly visible location.

Having conceded the principle that estates should be designed around the needs of vehicles, DB32 attempts to reduce speeds on the lowest roads in the hierarchy by short lengths of cul de sac, shared surfaces, introduction of bends and 'pinch points' and careful planting and landscaping.

However, it can still be seen as designing estates around the needs of the car. The central feature of traffic calming is that estates should be designed around the needs of the people, a principle that should be applied equally to shopping, business and even industrial areas.

The origins of modern ideas of traffic calming are generally acknowledged to have been in the Netherlands in the 1970s with the development of 'Woonerfen'. The basic principle is that, far from segregating traffic, it should be integrated but on the basis of priority being given to the needs of the people, rather than the needs of the traffic. Thus speed is strictly controlled, not only by regulation, but by the physical layout of the street. The idea of a central section, the carriageway, where traffic has priority is removed, and, instead, the traffic has to take its turn with other road users in manoeuvring past whatever other functions of the street get in the way. The environment is designed for people, with seats and trees and play areas, rather than designed for cars with a forest of traffic signs.

The origin of the phrase 'traffic calming' actually comes from the German 'Verkehrsberuhigung'. Around 1980, the Germans began to adopt the ideas that were being applied by their neighbours in Holland. Initially they combined ideas of the Woonerfen with the idea of environmental areas defined by street closures drawn from the British report on Traffic in Towns, but this has a number of disadvantages. Woonerfen are very expensive to apply in existing streets because of the need for large scale reconstruction of the pavement. The principle depends on the transfer of traffic to roads having higher capacities, but these roads are frequently still residential, so advantages for residents in the newly quiet streets are merely paid for by residents in streets where the traffic is concentrated. And vehicles following tortuous routes use more fuel and create more pollution. (See Figure 24.1.)

Road closures have been installed in a number of British local authority areas in order to divert traffic from residential streets onto distributor roads, more able to cope with it. This has significantly improved conditions in those streets where through traffic has been removed but it ignores the problem faced by residents and other users on multi-functional roads. (See Figure 24.2.)

The German idea, which has yet to be accepted in Britain, is to apply the principles of Woonerfen to existing streets, and to distribute traffic through residential areas so that it is not concentrated on particular routes, but to do so in such a way as to give priority to pedestrians, cyclists, or children playing without actually prohibiting people from reaching their homes in their cars or having their goods delivered by van.

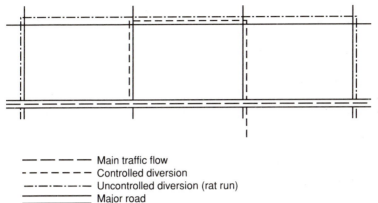

- — — — — Main traffic flow
- – – – – – Controlled diversion
- —·—·—·— Uncontrolled diversion (rat run)
═══════ Major road
─────── Minor road

Fig. 24.1 Diversion of traffic to roads more able to cope

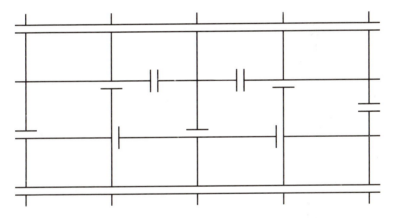

Fig. 24.2 Diversion of traffic by road closures

Controlling speed: British practice

It is generally accepted that something more than a legal speed limit is required if driver behaviour is to be significantly affected. Traffic engineers face something of a dilemma: speed limits, and their enforcement, tend to be resented by motorists. The police are unhappy about enforcing 30 and 40 mph limits, partly because they have better things to do and partly because it can undermine their position of mutual respect with members of the public who would not dream of breaking 'proper' laws.

On the other hand, recent research has shown widespread support amongst motorists for traffic calming measures which control speeds 'where people live'.

Driver behaviour is largely conditioned by the driver's perception of what is safe. His perception may be rather optimistic, but not wildly so. Ignoring the small proportion of completely irresponsible drivers who insist on treating the motorway as if it was Brands Hatch, or completely ignore weather, visibility or traffic conditions, and the equally small proportion of, usually juvenile, 'joyriders', we find that most people drive a little bit faster than perhaps they should.

Inter-urban roads are designed on this basis.[6,7] The 'Design Speed', on which geometric features like curvature, sight distance and superelevation are based, is intended to represent the 85th percentile speed of traffic using the road. It has been found that, on roads fully designed to the standards adopted in 1981, this in fact happens. But what also happens is that the 99th percentile speed is one design speed band above the design speed. The designer will work on the assumption that 99% of traffic will not travel faster than one speed band above the design speed, and although this may lead to a certain amount of passenger discomfort in the fastest vehicles, it will not lead to an unacceptable level of risk of collision or loss of control.

The author has undertaken practical tests of this hypothesis and found it works in reasonably standard situations. Using final year BEng students from Sheffield City Polytechnic, we carried out a series of speed measurements on sections of single carriageway, newly constructed highway having a design speed of 85 km/h. We duly found that the 85th percentile speed was 85 +/− 2 km/h and the 99th percentile speed was 100 km/h +/− 2 km/h, one design speed band higher. At one site, we found that the hypothesis did not work. This was the Stocksbridge By-Pass in South Yorkshire, where major variations were noted in speed of traffic travelling in different directions, but this was put down to the severe gradients which are a feature of that particular road.

In all cases, the legal speed limit was 60 mph, but this seemed to have little effect. It is the physical features of the road that determine driver behaviour.

The same appears to be true in urban areas. If the driver sees a long straight section of road ahead of him, with no obvious interruptions to traffic flow, an 85th percentile speed of 50 mph and 99th percentile speed of 60 mph plus is quite likely, even if the road is subject to a 30 mph speed limit. An unofficial alliance between motorists tends to develop, whereby a motorist passing a speed trap is likely to flash his headlights to drivers coming the other way to warn them to slow down. As soon as they've passed the waiting policeman, they speed up again. Sometimes the police quite deliberately place themselves in a highly visible position, not, particularly, to increase the number of convictions they can obtain but to cause a general slow down in vehicle speeds. Notices warning that a particular location is liable to have speed traps are posted for the same reason, and one police force has recently resorted to placing cardboard cut outs of patrol cars on motorway bridges.

So if our purpose is to reduce vehicle speeds in residential streets to between 6 and 20 mph, we must do something about the physical layout of the street. This is traffic calming.

Apart from the imposition of speed limits, low cost physical alterations that can be undertaken fall into the following categories:

1. Road surface measures, such as humps, tables and alteration of the surfacing material.
2. Road narrowing by means of alterations to kerb line to provide pinch points or throttles; white lining; arrangement of parking or planting.
3. Introduction of bends by means of chicanes; arrangement of parking or planting.
4. Treatment of junctions to provide minimal turning radii or mini roundabouts.

Of these, the most common introduced to date are road surface measures, particularly humps.

Although there are an increasing number of examples of the introduction of traffic calming measures on individual sites to solve particular problems – particularly sites with a bad accident record where vehicle speed has been identified as a factor in a substantial proportion of accidents – there is also increasing interest in the idea of area wide traffic calming measures. Administrative systems tend to discriminate in favour of those schemes where potential for preventing accidents can be demonstrated, as these are more likely to attract grant aid from central government. This potential is likely to be greater if an identifiable area, rather than a series of scattered sites, can be considered.

Road humps

Road humps are short, raised areas of the carriageway surface constructed in accordance with regulations laid down by the Department of Transport.[8] A glossy leaflet is available from the Department giving details and illustrations of the implementation of the regulations.[9]

Either round or flat topped humps, as shown in Figure 24.3, are permissible within the regulations. They will not cross the whole width of the carriageway as it is necessary to allow a minimum 200 mm wide drainage channel past the humps.

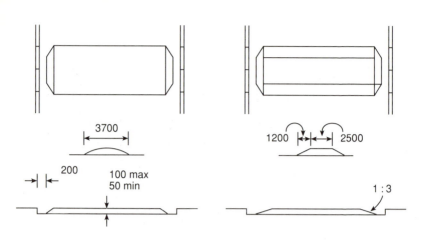

Fig. 24.3   Road humps

The most important restriction on the use of road humps is that they can only be used on roads subject to a 30 mph speed limit or lower. Humps may not be used on a trunk road, a special road or a principal road. They must always be preceded by a speed reducing measure, which may be another hump, certain types of junctions, certain types of road markings, certain types of bend and the end of a cul-de-sac.

There is no restriction on the use of humps on a bus route, although it may be considered desirable to use humps of less than the maximum height of 100 mm in order to minimise passenger discomfort.

Although road humps are now becoming an accepted feature of the highway, it is nevertheless important to undertake a thorough consultation exercise prior to implementation of a scheme. Those who might have their doubts about the benefits include the emergency services, public transport operators and businesses with premises affected. Occupiers on adjacent streets, to which traffic could conceivably divert in order to avoid the road humps, may also be concerned.

Humps can either be manufactured out of *in situ* materials, i.e. concrete with bituminous surfacing, concrete blocks or granite setts, or they can be obtained in preformed units from a variety of manufacturers.

An alternative to the use of humps is the use of tables, which consist of a raised area of carriageway several metres in length, frequently used at junctions so as to provide a 'hump' on both directions across the junction. This may be particularly appropriate in old housing estates laid out to a grid pattern where there are a large number of crossroads, which are potential danger spots.

Fig. 24.4 Traffic calming by variable surfacing

Fig. 24.5 Traffic calming by use of islands

Simply altering the surface materials is unlikely to be very effective as an alternative to humps and tables, but may have applications in the context of overall environmental improvements. The use of granite setts to delineate vehicle paths across a large paved area predominantly reserved for pedestrians can be appropriate in both environmental and traffic management terms. Figure 24.4 shows just such an application.

Other speed reducing measures

Various mechanisms of narrowing the road are available, involving the use of central traffic islands, chicanes, arrangements for parking, realigning the kerbs or simple white lining. One of these may be appropriate on main roads where humps would be prohibited. Examples are shown in Figures 24.5, 24.6 and 24.7

Road narrowing can frequently be implemented at junctions where

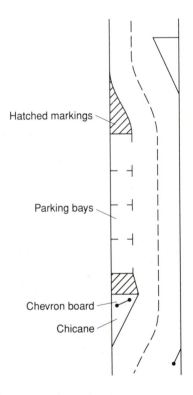

Fig. 24.6 Traffic calming by chicanes

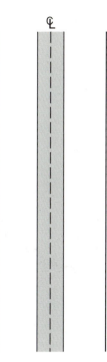

Fig. 24.7   Road narrowing by white lining

it is desired to limit speed of vehicles entering a residential area, as shown in Figure 24.8. Possible treatment includes restricting the radius of the kerb at the junction to about 1.0 m, providing a pedestrian way across the junction defined by different surfacing – e.g. concrete blocks in a contrasting colour, or, in certain circumstances, providing a mini-roundabout. This is likely to be expensive and may require additional road space, but where it is desired to slow speeds on all approaches it may have applications.

With the success of traffic calming measures developed and applied overseas, the wider application in Britain gradually became an objective for local authority traffic planners. Somewhat reluctantly, the Department of Transport came to support the idea of area wide traffic calming, both in the interests of reducing the number or severity of road accidents and contributing to the general environmental improvement of urban areas. In December 1990 the Department issued guidance to local authorities seeking to implement 20 mph speed limit zones[10,11] and established a series of trial areas to test their effectiveness.

The first of the trial areas to be implemented was at Tinsley, Sheffield, and this is more fully described in the case study shown in

Fig. 24.8 Road narrowing by planting

Fig. 24.9 Area wide traffic calming

Box 24.1. The most important point about the zones is that the speed limit should be largely self-enforcing. It is a system of area wide traffic calming involving the application of some or all of the techniques listed above at appropriate points. In the Tinsley case, the starting point was a Housing Action Area, so traffic calming was seen as a means of helping to improve the environment of a rather rundown community, and the 20 mph limit seen as a means of assisting, not enforcing, the attainment of these objectives.

Box 24.1:

Case Study: The implementation of traffic calming in a 20 mph zone. Tinsley, Sheffield, South Yorkshire.

The central objective of the Tinsley scheme, like others, is to influence the behaviour of drivers in such a way as to reduce conflict between them and other road users.

Tinsley is located on the side of the lower Don Valley close to the M1 motorway and the administrative boundary between the Sheffield and Rotherham Metropolitan Districts. Traditionally, it provided housing for industrial workers employed in the heavy steel industry of the valley. The area is made up of terraced housing facing directly onto streets laid out in a grid pattern. There is no off-street parking. It has Housing Action Area status and funds were available for streetworks as part of the general environmental improvement. There were significant numbers of child pedestrian accidents which were concentrated on roads which were bus routes and carried significant through traffic and had a considerable amount of parking.

The area was defined as that being in need of environmental improvements as part of the Housing Action Area proposals. Each entry point to this area was treated by means of carriageway narrowing, a flat topped hump and entry signs. A table has been constructed at each of the junctions within the zone, and the carriageway has been narrowed. This meant shorter crossing places for pedestrians, and the position was further improved by use of low kerb upstands to assist people with pushchairs, wheelchairs and trolleys.

On-street parking has been controlled by the areas of footway widening. On long straight stretches of road there are intermediate humps.

The height of the humps has been limited to 75 mm on bus routes and 100 mm elsewhere. Phase I of the scheme was completed in 1990 and phase II in 1992. The scheme has been monitored by the Transport and Road Research Laboratory, who have found that speeds through the area have been reduced from 25–40 mph to 13–20 mph. No injury accidents have occurred between the official opening of the 20 mph zone and April 1992.

Sheffield City Council took considerable trouble to involve the residents of the area in the planning process. In accordance with

Box 24.1 continued:

> Department of Transport regulations the order creating the 20 mph zone is temporary but will be made permanent when the Council can establish that there has been:
>
> 1. a significant reduction in speed of traffic travelling through the area;
> 2. a significant drop in the number of accidents.
>
> Sheffield City Council are satisfied that both these conditions will be met.

It is likely that traffic calming schemes will become steadily more common in the future. They are consistent with the Government's stated policy of substantially reducing the toll of road accidents before the end of the century. Implementation of traffic calming measures is no longer a question of whether, but rather one of when? Many other factors in transport planning, crucially those aimed at reducing congestion in city centres, will have an impact on implementation of traffic calming. Somehow, we have got to find ways of reducing dependence on the private car, not only because cities cannot, physically, be adapted to cope with it in the numbers likely to be seen, but also because of the whole range of environmental improvements that could follow.

References and further reading — Chapter 24

1. Tolley R. *Calming Traffic in Residential Areas.* Brefi Press, Tregaren, Dyfed. 1990.
2. Tripp H A. *Town Planning and Road Traffic.* Edward Arnold. London. 1942. Quoted in Tolley R.
3. Buchanan C. *Traffic in Towns.* London. HMSO. 1963.
4. Tolley R. op cit. p. 18.
5. Department of the Environment. *Guide to the layout of estate roads and footpaths.* Design Bulletin 32. HMSO. London. 1992.
6. Department of Transport Standard TD9/81. Highway Link Design, as amended. Reprinted 1990. (Under revision.)
7. Department of Transport Advice Note TA43/84. Highway Link Design. Reprinted 1990.
8. The Highways (Road Humps) Regulations 1990.
9. Department of Transport. Traffic Advisory Unit. Leaflet 2/90. *Speed Control Humps.* 1990.

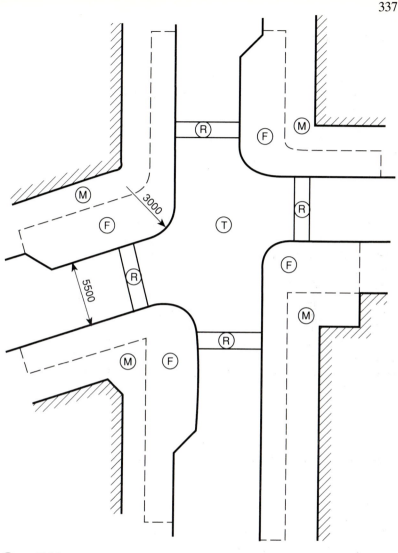

(T) — Table
Bituminous macadam regulating course on existing c/way surface
40 dense bituminous macadam wearing course

(R) — Ramp 75 high
80 block paviors on 50 sand bed
Existing c/way broken out as required

(F) — Footway
pcc paving slabs on 25 sand bed
150 Type 1 sub-base

(M) — Areas treated to tie in with existing levels

Fig. 24.10   Junction layout at Tinsley 20 mph zone.

338

10. Department of Transport. Circular Roads 4/90. *Basic Guidelines for the introduction of 20 mph speed limit zones.* December 1990.
11. Department of Transport – Traffic Advisory Unit. Leaflet 7/91: *20 mph speed limit zones.* May 1991.

PART VI

SCHEME APPRAISAL

Chapter 25

Cost Benefit Analysis

The basis of cost benefit analysis

It is one of the fundamental truths of capitalist economics that there are never enough resources to meet demand. This follows from the philosophy that says that human beings are basically greedy: however much we have, we will always want more.

This may seem like being on a rather high philosophical plane but it does fit in with our experience in the transport sector. Transport is rarely an end in itself: it comes about because we want greater choice in the jobs we can do or the goods we can buy or the places we can go on holiday. Even if we do not want it for ourselves, we want it for our children or our friends or our neighbours. The search for a 'better life' seems to be a constant part of the human condition, and the better life usually involves transport.

If demand can never be fully satisfied, it follows that there must be some mechanism for determining priorities. In the private sector, the market operates in such a way as to perform this function. However, for a market to exist there are certain preconditions: there must be a buyer and a seller and a product. The product must be clearly definable, and it must be the property of the seller at the time that the sale takes place. It is not necessary for it to be a physical object: a legal contract, for instance, can be sold, but it may only be a verbal agreement between two parties. For a sale to take place, the seller must transfer to the buyer the product and the buyer must transfer to the seller the value of the product, usually, but not necessarily, in money.

There are plenty of circumstances in the transport field where sales take place and a true market exists, at least within the limitations of government intervention, but these do not include the provision of roads which are free to the user at the point of consumption. In determining priorities for investm nt in the road system, we need an alternative to the operation of the market. The method adopted is a mixture of political procedures and economic analysis.

In the first instance, the decision to build a road is initiated by Parliament, the local highway authority or a developer. Details of the procedures involved are shown in Chapter 8. They involve the application of cost benefit analysis in the form of the Department of Transport computer progam, COBA,[1,2] and the development of frameworks by which the advantages and disadvantages of different routes can be compared.

It is reasonable to say that there are certain identifiable costs and benefits associated with the construction of a new road. An initial assumption is made to the effect that the total cost and benefit is equal to the sum of the costs and benefits to each individual. If these costs and benefits can be expressed in terms of a single unit, then a simple arithmetic calculation is all that is needed to determine the ratio between them. If, however, they cannot be so expressed, comparison is a more complex process. Expressing them in a standard tabular form, i.e. a framework, may be an aid to making a decision.

COBA

COBA9 is the Department of Transport's program for determining the scale of those costs and benefits which are measurable and convertible to monetary units, discounting them over the lifespan of the road, and calculating a ratio of benefits to costs. It is an essential part of the appraisal process and no scheme is likely to proceed unless it can be demonstrated that the benefit/cost ratio, as measured by COBA, is at least positive and greater than the minimum being implemented, for the time being, by central government. The COBA rate of return is one input to the frameworks, which are described in some detail in Chapter 8 in the context of public consultation.

The costs involved are:

1. The cost of the land taken into the highway.
2. The cost of designing and building the highway, including administrative and legal costs.

3. The cost of maintenance, policing and other ongoing costs.

The benefits involved are:

1. Savings in time for road users.
2. Savings in vehicle operating costs for road users, including reductions in fuel consumption.
3. Savings resulting from a reduction in road accidents.

Each alternative proposal for a scheme (a 'do something' option) is compared with the 'do minimum' option and the net costs and benefits provide the input to COBA. Thus, if a choice is being made between a Green Route, a Red Route and neither, then the costs and benefits of neither would be subtracted from each of the red and green values before computation. The present value (PV) of the costs is subtracted from the PV of the benefits for each alternative, to give the Net Present Value (NPV) of the project. This is divided by the initial capital cost of the scheme and expressed as a percentage to give the COBA rate of return. The priority of the scheme is determined by its position in a league table of rates of return.

The use of PVs is standard accounting practice to take account of the fact that money now is worth more than the same amount of money in the future. If you are given the choice between having £100 now or £102 next year, you would be sensible to take the first because you could invest it, very safely, at interest of, say 5%, and you would have £105 next year. If the choice was £100 now or £110 next year, the second might be the better bet for you might find it difficult to get a 10% return without taking significant risks.

So if money now is worth more than the same amount of money in the future, it follows that money in the future is worth less than the same amount of money now. Hence, the term Discounted Cash Flow (DCF) and the use of a discount factor. The amount by which values should be discounted depends upon the general level of interest rates, but these vary substantially and frequently. A standard discount rate is required if alternative designs for the same scheme are to be compared, and 8% is currently used for this purpose.[1]

COBA9 undertakes a very similar procedure to that shown for private investment in Box 25.1. Alternative designs of the same scheme will have different cost and benefit profiles so it is possible to use the COBA rate of return to compare them and decide which, on econmic grounds, represents the best value for money. It is also possible to compare all the schemes with the situation in which no new road is built – the 'do minimum' case – and hence decide whether, economically, we should leave things as they are.

COBA9 is not the only tool used in appraisal of schemes. There are other inputs to the framework and it may be that, in particular circumstances, something other than the scheme giving the highest rate of return goes forward.

Box 25.1:

The application of discounted cash flow techniques[3]

A very similar process to that included in COBA is used to assess priorities for capital investment in a free market. A contractor might, for instance, be considering the purchase of a new item of plant for £20,000. He might reckon that, after allowing for maintenance, fuel, drivers' wages and overheads, the item will earn him £5000 per year and after four years it will have a second hand value of £6000. He can draw up a table that looks something like this:

Year	Costs	Income	Value
0	20,000	5000	
1		5000	
2		5000	
3		5000	6000

In order to discount these figures at, say 8%, he needs to work out the discount factor, which is based on the reverse of the compound interest formula:

$$DF_n = 1/(1 + r)^n \qquad [25.1]$$

and the discounted cost or benefit is given by

$$PV_n = S_n \times DF_n \qquad [25.2]$$

where DF_n = Discount factor for year n; r = Discount rate; S_n = Value of cost or benefit in year n; PV_n = Present value of S_n.

We can therefore complete our table (using r = 8% = 0.08):

Year	Costs	Income	Value	DF_n	S_{cost}	S_{bens}	S_{val}
0	20,000	5000		1.000	20,000	5,000	
1		5000		0.926		4,630	
2		5000		0.857		4,285	
3		5000	6000	0.794		3,970	4764
					20,000	17,885	4764

The total costs of this operation are therefore £20,000, and the total 'benefits' £17,885 + £4764 = £22,649. Subtracting costs from benefits we get £2649 and dividing by the capital outlay we get 0.123, or a 'return' of 12.3%. The contractor may very well consider that this is a worthwhile investment and go ahead.

The problem with COBA is not in the calculation of DCF – this is a standard procedure with many other applications. It is in forecasting the size of the costs and benefits and assigning a value to them, and, hence, in determining the weighting that is given to each.

In the case shown in Box 25.1, that of a private company making a decision to invest in an item of plant, the profile of income and expenditure can be readily expressed in monetary terms. The objective of the company can only be to make a return on investment for its shareholders. Its costs are all made in the form of payments from its accounts and the only benefit that it receives is in the form of money paid in ('sales'). Even if transactions can be made in other forms, these transactions have to be valued and included in the annual accounts which are legally required by the Registrar of Companies, the Inland Revenue and Customs and Excise VAT officers.

The statutory nature of private companies is such that they are able to take risks. The shareholders of the company appoint Directors to administer the affairs of the company. There are a number of constraints on the freedom of action of the Directors, but, in general terms, they are free to manage the company's affairs on a day-to-day basis and endeavour to do so in such a way as to maximise the shareholders' return. There is no guarantee, however, that they will be successful. If a company makes no profit, then it is likely that the shareholders will receive no dividend. They may, in such circumstances, resolve to remove the Directors and appoint somebody else: that is their right. In more dire circumstances, in which the company makes a loss which exceeds the total of its assets, it may go out of business, in which case the shareholders may have lost the money they invested in the company: that is their risk.

Neither of these conditions apply in the case of the construction of a new road. Both costs and benefits accrue to society as a whole and are not easily expressed in monetary terms. Roads are, nearly always, the responsibility of a public sector bureaucratic body whose objectives do not include taking risks. The cost of a new road is made up of three elements: land, construction and maintenance.

The cost of the land is taken as the price that the highway authority will have to pay in order to buy it from existing owners, but this is very variable and the presence of the road can affect the value of the land. A 'best estimate' based on land use and value statistics has to be used.

The cost of construction is intended to represent the sum that will have to be paid to contractors and consultants for carrying out all the work involved in design and construction, but this can also be very variable. The Department of Transport uses competitive tenders wherever possible and can base its estimates on past data collected from similar schemes. But tenders are always likely to reflect the state of the construction industry at any given time, and there can be variations of as much as 30% depending on whether there is a glut of work around, in which case contractors can build up healthy profits, or whether there is a shortage, in which case they may be cutting their margins to the bone simply in order to survive. The Department is continuously experimenting with different forms of contract, including design and build, time and bonus/penalty systems and lane rental, with the aim of finding the best way of ensuring value for money without sacrificing quality or safety.

Maintenance costs are perhaps the easiest to predict in the short term, although there is considerable uncertainty looking further into the future, and standard assessments are carried out for a 30 year period from the time of opening of the road. They are also dependent on weather conditions, although these may average out over a long period. Allowance is made for the savings in maintenance expenditure on existing roads if traffic is diverted to a new road.

It is on the benefit side of the equation that uncertainty becomes most difficult to predict.

It is implicit in COBA that relative weightings must be applied to the value of time, vehicle operating costs, and road accidents in order to come to a single value of the rate of return. This is done by applying numeric values in money terms to each of the component parts of the benefits profile. By varying the relative values one can vary the value that comes out at the end.

The largest part of the value of the benefits within COBA comes from savings in journey time. Taking a relatively simple example of a new length of rural road being constructed to divert traffic away from an existing congested or tortuous section, it is clear that savings in journey time should result because vehicles will travel faster. The amount of time saved can be determined by measuring journey times on the existing road and making assumptions, based on fairly reliable data, about the relationship between design speed and actual speed on the new road. The total amount of time saved is taken as the sum of the individual time saved on each journey, so the distribution of different types of vehicle is likely to be relevant: cars travel faster than lorries on similar sections of road and, on two lane roads, are more likely to be able to overtake.

COBA operates by building up a fixed trip matrix for each year and hence determining time and vehicle operating cost savings for each link and each junction. These are then added to other benefits to give the total benefit for each year. A saving of one second by each of 100,000 vehicles equals a saving of 100,000 vehicle-seconds. So does a saving of 10 seconds by each of 10,000 vehicles or 100 seconds by each of 1,000 vehicles. A driver would be unlikely to notice a saving of one second in his journey time and would therefore value it at zero, but he might well ascribe a definite positive value to a saving of 100 seconds. This is to say that it may not be correct to ascribe, as COBA does, a straight line relationship between the value of time and the amount of time saved. The counter argument is that schemes should not be seen in isolation from other schemes. The total amount of time saved along a route, which may involve several projects, will always be significant.

It is still necessary to ascribe values, in terms of pounds and pence, to the amount of time saved by a particular improvement, or a particular design. There is no single figure that can be attached to a time saving by a vehicle. The benefit is not saving in the vehicle's time, it is saving of the occupants' time. So the total benefit figure for time savings is achieved by summing the value of time savings by all vehicle occupants.

Individual time savings, in pence per minute, are ascribed to different categories of road user, e.g. car driver, car passenger, public service vehicle passenger or goods vehicle driver. Survey data can be used to determine average vehicle occupancy, from which the number of road users can be determined from a simple classified vehicle count. Survey data can also be used to determine values of time for each category of road user. The values used are updated from time to time and, for example, are shown in Table 25.1.

COBA also contains a computation for determining vehicle operating costs based on vehicle type, link speed and gradient. Link speed is, in turn, determined from highway geometry and the assumption that free flow conditions exist. Loadings can be applied at junctions. Total vehicle operating costs include fuel consumption, which is likely to be greater at low speeds; wear and tear and replacement of consumable items (tyres, oil, brake linings) which is considered as a fixed rate per kilometre; maintenance, which is considered to vary with vehicle speed in the same way as fuel consumption; and depreciation, which is considered as a rate per kilometre on the grounds that cars are likely to have a life of between 70,000 and 100,000 miles rather than a fixed number of years, and similarly other vehicles.

Taxes (VAT, licence duty and petrol duty) are not considered to be part of vehicle operating costs as they constitute a transfer between the individual and the state. Summing the costs and benefits to all parties will mean that a tax is a cost to the individual but an identical benefit to the state. They will therefore cancel each other out.

Savings in accidents are particularly difficult to assess, as an attempt to place a value on 'what you would pay to avoid an accident' is impossibly subjective. Most of us would probably be prepared to pay a very high sum indeed if we could, by that means, ensure that we were not involved in an accident. We would probably pay an even

Table 25.1   Value of time per vehicle. Source: COBA9 Manual (1989 update)

Type of vehicle	Value of time/vehicle (p/hr) (1988 prices)
Working car	990.8
Non-working car	383.9
Average car	468.8
Light goods vehicle	859.0
Other goods vehicle	622.5
Public service vehicle	3213.7

higher sum to avoid an accident to our nearest and dearest, and probably a rather lower one for someone who lives down the street and a really low one for someone of whom we have never heard.

Accidents are classified as Fatal, Serious Injury, Slight Injury and Damage Only. The costs are assumed to have three components:

1. Loss of output.
2. Medical and emergency services, administration and property damage.
3. Costs of grief, pain and suffering.

Of these, only the second is relatively easily valued, although since there is no legal requirement to report a 'damage only' accident to the police, the figures are inevitably based on incomplete data. Statistics are available to determine the likely output of an average worker for the remainder of his working life, although whether it is reasonable to count this as a 'cost' in times of high unemployment is very debatable. A disproportionate number of casualties in road accidents are elderly people who have completed their working life and therefore have no future output. Placing too high a value on item (1) leads to the untenable conclusion that, for that part of the population past retirement age who form a large proportion of the accident casualties, society as a whole is actually better off without them.

Item (3) is intended to rectify this conclusion, although even here, there are problems. Arguably the worst, or, at least, the most expensive, type of accident, is the one in which somebody relatively young is seriously injured, leading to long term hospitalisation. The death of a victim is, by comparison, short-term and comparatively low cost. But it would hardly be acceptable for the government to adopt a policy which says that it was placing a higher value on avoidance of injury than on avoidance of death.

In order to apply the values of time, vehicle operating costs and accidents shown in the various tables taken from the COBA manual, it is necessary to determine the numbers of each category of vehicle that will use each link on the network being assessed throughout the period of the assessment (30 years from the date of construction). COBA does this, starting from a matrix of origin and destination data gathered from surveys. The principles of forecasting future levels of traffic growth are discussed more fully in Chapter 7 but they apply equally here. In the majority of cases, the road network that is likely to be affected by an improvement scheme will be more complex than simply replacing one section of road with another. COBA will develop a future fixed trip matrix for a comparatively simple network from existing flows, National Road Traffic Forecasts and any other data available.

COBA is not able to cope satisfactorily with complex networks in urban areas, nor where congestion is likely to occur on links which affect the proposals, as these will affect the traffic assignment which is implicit in COBA. More sophisticated models are being developed to represent urban traffic flows on complex networks and the Department of Transport has developed UREKA[4] in order to allow the output of these models to be used directly in the appraisal and satisfy some of the problems identified by the Standing Committee on Trunk Road Assessment in its 1986 Review of Urban Appraisal Methods.[5] However, it is unlikely that UREKA will, by itself, be enough to satisfy the critics. Their position is that it is absurd to attempt to model very complex relationships between traffic flows far into the future with any degree of accuracy when many of the variables that will reflect those relationships are simply not known.

References and further reading — Chapter 25

1. Department of Transport. *The role of investment appraisal in road and rail transport.* DTp. London. 1991.
2. Department of Transport. COBA9 Manual. DTp. London. 1982 and subsequent amendments/updates.
3. Institution of Civil Engineers. *An Introduction to Economics for Engineers.* ICE. London. 1962.
4. UREKA – has the DTp found the answer to COBA's problems? Local Transport Today. 27 November 1991.
5. Standing Committee on Trunk Road Appraisal (SACTRA). *Review of Urban Appraisal Methods.* HMSO. 1986.

Chapter 26

Return on Investment Criteria

Charges to the customer

Many parts of the transport system, other than roads, are in the private sector and thus investment is justified if it can be recouped through charges. It is the intention of the Government to move further down the road of privatisation with the recently published proposals for the railways. It will be a matter for the private companies in whom ownership is vested to make their own decisions regarding investment, raise their own capital and implement plans in their own way.

In the meantime, however, substantial parts of the railway will remain in public hands. The procedures for authorising investment follow a parallel path to that which would be followed by a private company. British Rail, in order to justify capital expenditure, must forecast the amount of that expenditure and the amount of revenue generated through payment of fares and guarantee a return on the investment of 8%. This is necessary to obtain government sanction for carrying out the work, although British Rail still has to raise the necessary money from its own resources, i.e. by charging the customers.

This system of appraisal discriminates against rail schemes if a direct attempt is made to compare them with road schemes. Using cost benefit analysis, all savings in accidents, time and vehicle operating cost count towards the benefits accruing to the scheme. With a railway, however, these only count in so far as their value is reflected in the fares charged. No account is taken of the savings that accrue to other road users as a result of some travellers transferring to rail. Since one of the major questions facing planners in many cities is precisely the choice between a road option and a rail option, it is not very satisfactory to use two different sets of criteria for the comparison.

Justifying grant aid

Section 56 of the Transport Act 1968 empowers the Secretary of State to give grants to local authorities, British Rail and others towards the capital costs of improving public transport. A Section 56 grant has been a major factor in the funding of capital improvements to public transport which are currently being implemented or under consideration.

The promoter of a rail project is likely to find it necessary to put an individual case to the Department of Transport for Section 56 funding. The case put forward, which must include a full appraisal of the alternatives, is very much up to the promoter, and, indeed, may have to be changed in response to statutory changes introduced during preparation and the promoter's perception of which aspects will carry most weight.

The evaluation process used for Manchester Metrolink has been described by Tyson[1] and is contained in publications by Greater Manchester Passenger Transport Executive (GMPTE).[2,3]

In order to justify the scheme it was necessary for GMPTE to carry out a cost-benefit analysis comparing:

1. A base case in which existing rail services would be replaced by buses.
2. The retention and modernisation of existing rail services.
3. The introduction of LRT.

Costs that were evaluated as part of this procedure were:

- additional capital costs of construction and operation
- savings in capital costs for bus replacement
- operating costs
- savings in bus operation.

Benefits were:

- passenger time savings
- additional revenue.

The modelling process that was required to evaluate these benefits was extremely complicated, as it involved an estimate of the modal switch from bus to LRT and car to LRT and the summation of the time savings that would result. A major element of the time savings came from the penetration of the city centre by LRT vehicles, thus saving walking time from the two relatively inconvenient termini at Piccadilly and Victoria.

Approximately 54 urban rapid transit schemes are currently in the course of preparation, although many of these are at a very preliminary stage. Most would depend on Section 56 funding and this would be dependent upon a very marked improvement of the economic state of the country. Those currently under construction are Sheffield and the Beckton Extension to the Docklands Light Railway. Other schemes at a fairly advanced state of preparation are:

Bristol	– Parliamentary Bills withdrawn. Dependent on S56 finance.
Croydon	– Full private finance.
Glasgow	– Part of Strathclyde Regional Transport strategy. Requires S56 funding.
Leeds	– Parliamentary bill in progress. Requires S56 funding.
Nottingham	– Alignment safeguarded. Requires S56 funding.
West Midlands	– S56 application frozen by Government 1992.

In the field of LRT there have only been two substantial grants given under S56 – Manchester and Sheffield – although West Midlands has had some grant aid in order to carry out preparatory and design work. Given the uncertainty of government funds being available, proposers are now looking into the feasibility of increasing the proportion of funds that could be available from other sources, particularly the private sector in the form of design, build and operate concessions. This is the solution that was adopted in Manchester, with the successful contractor, a consortium of GEC–Alsthom, Mowlem and AMEC, being granted an operating licence for 15 years. Five consortia competed for the contract, but it was still felt that the S56 contribution was essential. If there is going to be much further progress in bringing LRT to British cities, it seems likely that private sector finance is going to have to meet most, if not all, the bills, and this is only likely to be available in return for long term operating concessions.

A number of schemes for restoring rail services outside city centres are also being prepared, and these, too, depend on S56 funding for a proportion of their costs.

The two most advanced proposals are the 'Ivanhoe Line' (Loughborough – Leicester – Ashby – Burton – Derby) and the 'Robin Hood Line' (Nottingham – Mansfield – Worksop – Retford). Both are being promoted by consortia of local authorities in the areas through which they pass, the lead being taken by Leicestershire County Council and Nottinghamshire County Council respectively. The viability of both schemes has been established to the satisfaction of the promoters and applications to the government for S56 grant and capital expenditure approvals are in the course of preparation.

The Ivanhoe Line involves reopening a disused British Rail freight line to passenger services and construction of new stations associated with park and ride facilities. The major civil engineering works will be at the stations. The new service would be extended at both ends to give local stopping trains on inter-city lines between Loughborough and Leicester and Burton and Derby. Since most of the track is already there, the heaviest item of expenditure would be on the trains, for which it is likely that some kind of leasing arrangement would be possible. The benefits which are required to justify the proposal in cost benefit terms include substantial journey time savings to non-rail users through reductions in congestion in the major towns involved.

The Nottingham scheme is slightly more complicated in that it involves reopening a length of tunnel that had been filled in following the original closure of the line. Apart from that, it involves opening freight lines to passenger traffic serving commuters travelling to Nottingham, restoring a rail service to Mansfield, the largest town in Britain without one, and providing a link northwards for the whole of the corridor, including Nottingham, to the East Coast Main Line at Retford.

The costs of the proposal are estimated at £16 million, to include new stations and car parking, Sprinter trains, track work and resignalling.[4] The line would generate at least £2 million per year in revenue when fully operational as well as providing substantial improvements to the environment through reductions in congestion in the Greater Nottingham area. Extensive research has been carried out into the proposals by the Institute for Transport Studies at the University of Leeds using stated preference techniques and this has concluded that up to 5000 people per day would use the line and that the northwards link is an important contributor to revenue. It is very difficult to establish a value for the benefits that would accrue by, for instance, reconnecting Mansfield to the rail network, but in the context of an expansion of rail freight facilities they would be considerable.

There are probably a number of other routes similar to Robin Hood and Ivanhoe which could be shown to be viable as public transport corridors if the initial capital involved in reopening them could be found. There is also potential for more rail services in tourist areas. Government proposals to privatise British Rail have brought expressions of interest from companies such as Stagecoach, which runs private rail services in Scotland as well as local bus and long distance coach services, Badgerline, which is a bus operator based in the south west and Richard Branson's Virgin Atlantic, which is an airline. It may be that availability of capital is not the problem that it might sometimes appear.

The major problem appears to be political. Those schemes which have progressed have had initial support from the local authorities

354

through whose areas they pass. Not all authorities are innovative or prepared to look further into the future than next year's budget. Departmentally, transportation is vested in the County Surveyor's Department or its successor, and the primary role of the County Surveyor is to build, maintain and manage roads. Present proposals for local government reform indicate 'hat county authorities will shortly cease to exist, and although this may mean a break with the past, it will also mean fragmentation.

So putting together a financial package that does not place more strain on the public purse is likely to be the major task facing those who would like to see, for environmental reasons, a greater role for public transport.

References and further reading — Chapter 26

1. Tyson W J *Planning and financing Manchester Metrolink*. Proceedings of the Institution of Civil Engineers: Transport. Paper 9835. 95. Aug: 141–150. August 1992.
2. Greater Manchester Passenger Transport Executive. Section 56 *Grant Application*, July 1985. GMPTE. Manchester. 1985.
3. Greater Manchester Passenger Transport Executive. Section 56 *Supplementary Submission*, March 1987. GMPTE. Manchester. 1987.
4. Nottinghamshire County Council. *The Robin Hood Line Project*. Report from the Director of Planning and Economic Development and the County Treasurer to the Policy and general Purposes Committee. Nottingham. 1991.

PART VII

THE FUTURE

Chapter 27

Where Do We Go From Here?

Is there a crisis?

Everybody will have their own opinion on whether the problems which were discussed at the 1992 World Environment Summit in Rio de Janeiro constituted the last hope for mankind and his polluted planet, or whether it was just another excuse for politicians to congratulate themselves on their own wisdom. Be that as it may, there are certainly opportunities for being less profligate with the world's non-renewable resources, and less ready to simply throw out our rubbish to pollute the rivers and the atmosphere. Many of these opportunities are in the field of transport, so as transport planners and engineers we are in a position to make a contribution.

Whether or not the inadequacies of our transport system are threatening to the planet, they certainly cause inconvenience. Congestion, which is now commonplace in urban centres through ever lengthening peak periods, is a major contributor to atmospheric pollution because engines which are ticking over and stopping and starting are less efficient than engines which are driving a car at constant, high speed along a motorway. But it is also extraordinarily unpleasant for those who are stuck in the jam and those who are walking past it or live beside it. And it is such a waste of valuable time!

So whether or not we have a world threatening crisis, it is essential that we solve, or at least reduce, the problem of urban congestion.

The primary cause of congestion, pollution and threat to life is our addiction to the private motor car and the very valuable role that is played in our economy by the lorry. Other modes have their problems, too, but they are minor by comparison.

Aeroplanes are major polluters in terms of noise and gaseous emissions and are very extravagant in the use of non-renewable fossil fuels. The harmful effects, particularly of noise, are concentrated around airports. Air travel is the safest method available for although occasional disasters become headline news, they are very, very rare. If, however, we take the fear of accident as well as the actual event into account, as it is suggested we should do when considering the impact of road traffic, the cost of flying might become higher. As with other forms of travel, the amount of demand tends to vary with economic performance. Assuming that the national economy grows, as measured by Gross Domestic Product, then it is reasonable to suppose that the demand for air travel will grow too: there will be more businessmen trading in America or Africa or China and more people who can afford holidays in Florida.

One target of the European Commission is to rationalise the market for air travel across the Community. Comparisons have been made between the high cost of fares in Europe compared with those in North America, and one reason for this has been the proliferation of small, national carriers supported, for reasons of prestige, by governments. Such restrictive practices are alien to the concept of the community and, sooner or later, will go. This will bring fares down, particularly on routes that are likely to have high levels of demand. Britain stands to gain out of these changes, partly because all international and some national routes involve sea crossings and do not have to compete with land routes (until the Channel Tunnel opens) and partly because our National Carrier, British Airways, is one of the largest and has had to cope with the disciplines of the free market for some time. Although some attempts to establish 'second force' airlines, most recently British Caledonian, have eventually been unsuccessful, there are a number of thriving private sector carriers like Virgin, British Midland, Air UK and Loganair coping with various sectors of the market. There will be real competition on routes into and out of Britain, and on internal routes involving sea crossings to Northern Ireland and the Scottish Isles.

As demand for air travel increases, so does congestion around airports in both airside and landside areas. Demand for international travel is dominated by London and Heathrow has had major problems of congestion for many years. The construction of Terminal Four relieved the problems of passing travellers through the airport but major problems are still faced in trying to reach the airport and by airlines in organising and sticking to their landing slots. There is no scope at Heathrow for increasing the number of aircraft that can be handled so intense efforts are made to get airlines to transfer their London flights to Gatwick or Stansted.

Provincial airports have the same problem to a lesser degree. Manchester has increased its runway capacity so as to incease the number of aircraft it can handle, built a new terminal to increase passenger throughput and the direct rail link is under construction so as to avoid the danger of missed flights because of hold-ups on the M56.

The most environmentally friendly mode of transport is undoubtedly by water. Ships can carry both goods and people efficiently and cause little pollution except in localised areas around ports. They are very safe, although the capsize of the cross channel ferry *Herald of Free Enterprise* at Zeebrugge raised concerns about the inherent safety of ro-ro ferries with huge open vehicle decks at or below the water line. More recently, the deaths of two children from asphixiation on the Irish ferry, Celtic Pride, sailing between Swansea and Cork, also raised concerns regarding safety. The prevention of toxic fumes entering passenger accommodation from the ship's sewage system should not be an insuperable problem. The establishment of watertight compartments on vehicle decks is probably impossible without a complete rethink of the design principles involved in getting vehicles on and off ferries very quickly.

Inland waterways have, in general, been allowed to decline in Britain, apart from those which are made up primarily of rivers. There is some goods traffic on the Thames and coastal shipping comes up the Humber as far as Goole. Some ship building takes place at Selby. Coal barges use the Aire and Calder Navigation, but attempts to attract traffic to the Sheffield and South Yorkshire Navigation, which has been upgraded as far as Rotherham, have largely failed.

Substantial works have been undertaken to make many of the old narrow canals usable by pleasure craft, including the Llangollen, Trent and Mersey, Kennet and Avon and Leeds and Liverpool. One of the most ambitious projects involves the Huddersfield Narrow Canal linking Huddersfield, West Yorkshire with Stalybridge, Greater Manchester and hence, the rest of the canal system. A lot of work has been done to dredge lengths of the canal that had become silted up and to restore locks to working order, but the major tasks still lie ahead. These include the restoration of the 3 mile long Standedge Tunnel under the Pennine watershed and reconstructing lengths of canal where it has been filled in and built over in Huddersfield and Stalybridge Town Centres, at Slaithwaite and at Mossley. It may seem churlish to seem critical of an essentially amateur organisation that is built on voluntary labour and under-funded government training schemes. If the Huddersfield Canal Society achieves its eventual objective it will have completed a multi-million pound project that many derided at the outset. But we should surely question whether inland waterways will ever again play the vital part in the transport network that they do in much of northern France, Belgium and the Netherlands if we continue to give priority to facilities that can only serve to allow people to engage in their passion for messing about in boats.

To some extent, the railways have suffered from the same British enthusiasm for all things amateur, although the professional railway has, at least, survived and is, once again, beginning to move forward. Railways, like canals, have a vital role to play in the future transport network, but it is less than helpful to meet the assumption, from the Department of Transport downwards, that the only way a railway can be reopened is if it is going to provide a mile or two of track on which to run gleaming and beautifully restored steam engines. Fortunately, local authorities like Nottinghamshire, Derbyshire, Strathclyde and Devon, as well as those in the metropolitan areas, are now beginning to recognise that rail based systems have considerable operational and environmental advantages in providing passenger transport in densely trafficked corridors.

Not that all our problems would go away if only everything went by rail. In environmental terms, rail has certain advantages over road transport, but only if it is used to something approaching capacity. Rail is compatible with electrification, which means that there is no pollution from the train. The electricity, however, has to be generated, so the pollution and consumption of fossil fuels takes place at the power station. The track has to be built and maintained, a process which also uses up energy and raw materials. The train consumes much the same amount whether it is full or empty. The advantage comes from the greater number of people or tonnage of goods that can be carried. If they are not being carried, there is no advantage.

Trains create noise which can be just as intrusive as noise from road traffic. Electrification of the power source, the use of continuously welded rail and the move away from marshalling individual wagons at different stages in the journey has greatly reduced the amount of noise generated by rail operations. Were there to be a move back towards the use of wagonload freight, which would be essential if there were to be a significant shift of goods traffic back on to rail, it would have to be done on the basis of goods wagons marshalled continuously into trains, possibly utilising containers or the type of ro-ro operation that will be used on the Channel Tunnel.

Given that the problems of urban areas are frequently associated with the use of private cars for journeys into the town centre, and that the reconstruction of town centres to cope with the demand is neither desirable nor practicable, then the only solution is for those journeys to be transferred to public transport. In some situations, there is scope for the construction of rail based LRT systems, but these are expensive and can only be justified on very heavily trafficked corridors. In most situations, the answer will be the use of buses. The implications of transferring a greater proportion of passenger travel to buses are considered below, but at this point it is important to consider what environmental advantages there are to bus travel. A bus, after all, is a large, heavy vehicle. It happens to carry people rather than goods, but otherwise it is very similar to a lorry. So its advantage comes solely from the fact that it can carry more people per unit of road space than a car. A full bus carrying 75 passengers will perform the same travel function as 62.5 averagely loaded cars. But an empty bus performs no function at all: it merely wastes fuel.

The future of road building

It seems to have become a truism to say that predicted growths in travel by car cannot be accommodated on our existing road system and the economic resources to expand the road system to cope with traffic demand will not be there. The problem is worse in urban areas, where expansion of the road system would mean unacceptable degrees of damage to the fabric of towns and cities, but in rural areas, too, concern is expressed about the growth of congestion.

This does not mean, however, that there is no future for the engineer specialising in construction of new highways, although maintenance and management are likely to be relatively more important than they have been in the past. However inconvenient it may be for those who would like to plan a car-less future, car ownership remains a high priority in most people's aspirations. Modern democratic government can only be achieved by the consent of the majority. The only alternatives are anarchy and dictatorship. In any event, many of the journeys that we would regard as essential cannot be undertaken by any means other than by car. The physical fabric of the society that we have created is such that many people depend upon a private car for many essential activities of their daily lives. That dependence cannot be changed in the short term, and even in the long term change will involve unacceptable costs to personal freedom and to the environment that we are trying to preserve.

The policy of the British Government is to make some moves in the direction of accommodating the forecast traffic growth without, if possible, alienating too many of its supporters. That is a logical reaction, perhaps the only possible reaction, by an elected government that can be removed from office if it loses popular consent, however that might be measured. So we have, in the roads programme announced in the 1989 White Paper, a £12 billion programme of improvements to the trunk road system. Priority will be given to widening motorways within existing corridors, and in providing by-passes where the environmental gains in the towns by-passed will be sufficient to justify the cost of the new road. Widening beyond the provision of a 4-lane dual carriageway has been found in the United States to be unsuccessful because the amount of weaving required by vehicles seeking to reach the outer lanes has tended to discourage such manoeuvres and to increase hazards, so where traffic levels appear to justify five or more lanes, as on parts of the M25, additional capacity is provided by constructing service roads. A ten lane route can be provided by four carriageways, dual 3-lane as on most existing sections of motorway and an additional dual 2-lane service road, one carriageway on each side of the main or existing route.

The roads programme as proposed has run into trouble on two counts. General economic performance has proved unable to provide funds for such a massive construction operation. The central policy of the Conservative Government is to control inflation at all costs. Inflation leads to escalating prices, loss of export markets and, in due course, rapidly rising unemployment. The Government is ideologically opposed to intervention in the market on the scale that would be practised by a Labour Government because it believes that such intervention is itself inflationary and therefore self-defeating. It is left with the very unpopular weapons of raising interest rates and cutting back on public expenditure. The optimism that greeted Mr Major's victory at the polls in April 1992 appears to have been unfounded so it seems likely that funds will not be available to construct new roads on the scale envisaged in 1989.

Even the policy of restricting operations to existing corridors has run into objections on environmental grounds. In particular, Surrey County Council, which is philosophically a supporter of the Conservative Government, is opposed to proposals to provide a service road alongside the M25 because of the environmental damage that it would cause to the area through which it passes, the concentration of damaging exhaust fumes and, crucially, the effect of generated traffic on side roads leading to the motorway.

The Government's policy of severely restricting expenditure by local authorities has always seemed at odds with its national transport policies. Only rarely does traffic begin and end its journey on a site immediately beside a trunk road. Indeed, the Department of Transport will restrict access to trunk roads in the interests of safety and maintaining free flow conditions. So every trip begins or ends on a local authority road. If traffic is to increase on trunk roads, then it will increase on local roads, too. To invest in one and not the other is quite illogical and will probably prove untenable. However large or small the total amount of investment in roads is to be, and that will be determined by central government, it has to be spread more equitably between local and national expenditure if the Government's own objectives of improving conditions where people live and work are to be achieved.

Integrating management of the environment

The title of this section is in fact what every politician aims to do. Many engineers are directly employed by the politicians, either as an employee or a consultant, so it is what we should aim to do as well.

It has been a recurring theme throughout this work that transport is rarely an end in itself. We travel because we need to get somewhere, in order that we can work, or go shopping, or conduct business or simply laze on the beach. The problem that is frequently coming to the top of the agenda is that, by travelling, we frequently impede the freedom of others through pollution, or congestion, or creating hazards or using up resources that are therefore not available to anybody else. And yet we claim to be concerned, in our democracy, to live in reasonable equilibrium with our neighbours and to pass on our facilities in good order to future generations.

As is described in detail in Chapter 24, schemes are frequently designed for the specific purpose of improving the environment in areas where people live. This is not new. The pedestrianisation of shopping streets during the 1970s was a process designed to remove the conflicts between different road users in order to promote a more pleasant, safer shopping environment. Initially, shopkeepers tended to oppose such developments but it rapidly became clear that the public tended to like them. Any loss of casual trade caused by the inability of a motorist to drive up and park immediately outside a shop was more than outweighed by the additional custom generated.

Such developments, however, tend to be designed in isolation from anything that is happening elsewhere in the neighbourhood. A central area precinct is, typically, surrounded by a ring road and a series of parking lots. Traffic calming schemes tend to be designed to overcome particular problems that have been identified, particularly those associated with the speed of traffic. No attempts have yet been made to integrate all aspects of traffic planning and development policy for the whole of an identifiable area, such as a town.

A greater role for public transport

One of the factors that discriminates against truly comprehensive transport planning is the separation of management structures for public transport from those for road traffic. This is an inevitable consequence of the deregulation of bus services introduced by the Government as part of its strategy to minimise monopolies throughout the public sector. As the White Paper on privatisation of the railways made clear, it is the Government's belief that the private sector is intrinsically more efficient than the public because it is subject to the disciplines of competition within the market and will attract entrepreneurial skills amongst its managers.

Overall, the results of bus deregulation were mixed. The total subsidies required were reduced because new, private operators found that they could run profitably on services that had previously been subsidised by the local authority. New services were introduced on routes that had previously been neglected because operators, faced with the need to find sources of income to replace the previous subsidies, went out to look for them. In many rural areas, particularly, new services sprang up where a local operator no longer had to compete with a heavily subsidised public company.

There is no evidence to suggest that safety levels have reduced despite repeated accusations from one side of the political spectrum. Drivers and vehicles are still subject to stringent tests in order to obtain licences to operate, and the danger of seriously underfunded maintenance facilities, apparent in some heavily subsidised public operations some years ago, has passed. There have been some much publicised cases of rival firms' drivers racing to collect passengers from a single stop and companies seeking to adjust timetables to their own advantage but such tactics must be shortlived. In the final analysis, public transport operates on a network system and some co-operation must exists between rivals if the public need is to be served.

It is not clear, however, that the present structure is the best one for maximising use of public transport. Private operators will aim to maximise profitability, and this is not necessarily done by maximising the number of passengers. Peak operations are, by their nature, unprofitable, because they utilise plant for a short period of time that stands idle for the rest of the day. So maximising profit may mean that the operator avoids providing a service at peak time. This is precisely the time when modal transfer from car to bus needs to take place if the worst effects of congestion are to be avoided.

As planning authorities, local authorities are bound to be making decisions which will have major implications for the profitability of transport operators, and will depend on those same operators for their implementation. In determining a Unitary Development Plan under the Town and Country Planning Act 1990, the local authority is obliged to take note of the requirements for transportation. That is sensible, for the location of a major housing area, employment area or shopping centre will have major implications for the scale of transport facility that is to be provided. But the amount of land taken up, and the nature of the development, will be very different depending on whether trips are to be predominantly undertaken by private car, by bus or by a fixed track facility such as LRT or heavy rail.

Structures to make it happen

The Government is committed to a major overhaul of the local government system in this country. The last reorganisation, brought about by the Local Government Act 1985, was regarded by central government as being successful. The abolition of the metropolitan county councils in England removed a tier of bureaucracy which the Government regarded as being a barrier to progress. The main argument against abolition was that the county councils had built up substantial areas of expertise that were required to undertake major development works, and that some functions, particularly transportation, could only operate on a county wide basis. However, the level of expertise required became a thing of the past with the transfer of the road construction units to private sector consultants and the effective ending of major road construction by local authorities. Passenger Transport Authorities have continued in existence as joint committees of district councils and there seems to be little evidence to suggest that they have not been as effective as their predecessors.

It is the Government's intention that local authorities should concentrate on becoming enablers of services rather than providers. Provision should be ensured by contracting out. To this end, the Government will move towards a system of unitary authorities for the whole of the country. These bodies will only employ, directly, a comparatively small number of people who will be primarily concerned with placing and supervising contractors. The size and boundaries of the new authorities have yet to be determined but the expectation is that they are more likely to be formed out of existing districts than out of existing counties. This has major implications for highways and transportation staff currently employed by county councils.

Some groups are already taking steps to establish in-house consultancies which can continue to bid for work after the abolition of the county council of which they are presently a part. This is a similar process to the reconstitution of direct labour organisations as in-house contractors following the introduction of compulsory competitive tendering for highway maintenance under the terms of the Local Government Planning and Land Act 1980 and the way in which refuse collection, sports facilities and refuse disposal have come to be organised. It is perhaps analogous to the management buy-out that sometimes occurs when a large company wishes to shed some part of its operation. We might therefore see, by the end of the century, a local government system based entirely on unitary authorities with populations ranging from around 100,000 to 750,000. These councils will not be the huge employers that they are today but will have small, highly qualified staff, many of whom will be concerned with policy development and strategic planning.

In the transportation field, a very large number of contracts will be administered. These might include the design and construction of one-off jobs, or they might be annual or tri-annual contracts for providing a particular service. The contractor might be a company we would recognise today (possibly a consulting engineer) or it might be a new form of organisation developed from an existing council department. It would provide its services in accordance with a defined contract which would, normally, be awarded on the basis of a competitive tender drawn up in response to a specification.

Whether this system would help or hinder the development of integrated transport systems remains to be seen. Some would argue that a free market, in itself, discriminates against comprehensive planning. But others would put forward the view that a comprehensive plan can only be implemented on the basis of maximum efficiency and that implies competition, not bureaucracy.

The transport of the future will evolve from the transport of today. There will be no big bang, after which everything is different.

Hopefully, we will see improvements in environmental quality brought about by technological improvements to vehicles and more successful traffic management. Restrictions on travel by car may be inevitable, but it remains to be seen whether this can be achieved by positive planning. It seems unlikely that there will be much reduction in the popular desire to own a car, so can we make progress by rendering some journeys unnecessary or can public transport be made so good that some people will leave their car at home some of the time?

Present trends suggest that lorries will become even more dominant in goods transport. Attempts to reverse the transfer of traffic from rail to road have not proved successful. Yet the scope for more rail involvement must be substantial given the labour intensive nature of road haulage, the spare capacity in the actual and potential rail network and the unpopularity of lorries.

In Chapter 1, it was stated that an efficient transport system is a prerequisite of industrial, commercial and social progress. That is just as true today as it ever was.

Appendix A

List of Central Government Offices relevant to transport

Departments of Transport and Environment, Headquarters:
2 Marsham Street, London SW1P 3EB
Scottish Office, Department of Environment (incl. Roads)
New St Andrews House, Edinburgh EH1 3DD
Welsh Office, Transport and Environment Sections
Government Buildings, Ty Glas Road, Llanishen, Cardiff CF4 5PL
Northern Ireland Office, Department of the Environment
Stormont, Belfast BT4 3SS
Regional Offices of the Departments of Environment and Transport:

Northern
Wellbar House, Gallowgate, Newcastle-upon-Tyne NE1 4TD
Yorkshire and Humberside
City House, New Station Street, Leeds LS1 4JD
North West
Sunley Towers, Piccadilly Plaza, Manchester M1 4BG
West Midlands
Fiveways Tower, Frederick Road, Edgbaston, Birmingham B15 1SJ
East Midlands
Cranbrook House, Cranbrook Street, Nottingham NG1 1EY
East
49/53 Goldington Road, Bedford MK40 3LL

South West
Tollgate House, Houlton Steet, Bristol BS2 9DJ
South East
Senet House, Station Road, Dorking, RH4 1HJ
375 Kensington High Street, London W14 8QH

Appendix B

Horizontal alignment example

Note

This example is concerned with the layout of a simple curve which joins two fixed straights. The curve has three elements: a circular arc and two transition spirals. The transitions are mirror images of each other.

More complex versions involving continuous transitions or reverse curves are possible but likely to require computer analysis.

Stage 1
Identify the design speed appropriate for the road.

Note

DTp Standard TD9/81 and Advice Note TA43/84 quote six standard design speeds: 120, 100, 85, 70, 60 and 50 km/h. Lower design speeds may be found on urban roads, but other constraints are likely to be more significant.

Choice of design speed follows from road type and function. On rural roads they are likely to be:

Motorways	120 km/h
Other dual c/ways	100 or 120 km/h
Single c/ways	85 or 100 km/h
Minor roads	70 km/h

For this example we will use a design speed of 100 km/h.

Stage 2
Identify the straights which are to be joined by the curve. These can be identified by co-ordinates, bearings or a combination of the two.

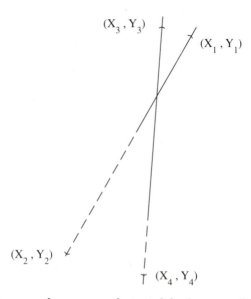

For this example, assume that straight 1 passes through (0, 0) and (40, 400) and straight 2 passes through (40, 400) and (300, 640).

We have assumed that both straights pass through the point (40, 400) i.e. we know the intersection point. We might have had to use less convenient points to define our straights but we could then have calculated the IP and, more importantly, the angle between the straights, the intersection angle, θ.

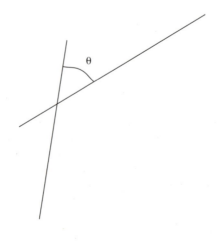

Stage 3
Calculate the intersection angle, θ

$$\text{Bearing of straight 1} = \tan^{-1}\frac{40 - 0}{400 - 0}$$

$$= 5°42'38''$$

$$\text{Bearing of straight 2} = \tan^{-1}\frac{300 - 40}{640 - 400}$$

$$= 47°17'26''$$

∴ θ = 41°34'48''

Stage 4
Select the radius of the circular arc that will form the central element of the curve.

Note

The minimum acceptable radii are shown in Table 3 of DTp Standard TD9/81.

Section B of this table refers to minimum radii of horizontal curvature. These are developed from calculations of superelevation, which must always be less than 7%. The table introduces the concept of 'desirable' and 'absolute' which gives the designer some leeway to choose circumstances when 'absolute' standards might be appropriate. There is no need to stick to the values shown in the table. Any value of radius greater than the minimum is acceptable.

For this example we will select 'desirable minimum' standards, i.e. Row B4.

Desirable minimum radius for 100 km/h = 720

Stage 5
Select rate of change of radial acceleration, q.

Note

Vehicles approaching a curve are going to experience a change of condition from a situation where radial acceleration = 0, i.e. the vehicle is proceeding in a straight line to a situation where the radial acceleration is some positive value appropriate to the radius of curvature. The length of road where this change can take place is called the transition. The transition curve is a spiral in which the rate of change of radius is constant. Rate of change of radial acceleration is proportional to rate of change of radius.

It has been found that the rate of change of radial acceleration for a road vehicle following a transitional path lies between 0.3 and 0.6 m/sec³. These can be considered 'desirable' and 'absolute' values respectively.

We are using desirable values so, for this example, we will take rate of change of radial acceleration $= 0.3$ m/sec^3

Stage 6
TD9/81 gives a formula for the calculation of the length of a transition. It is:

$$\text{Transition length } L = \frac{V^3}{46.7 \times q \times R}$$

where V = Design speed km/hr; q = Rate of change of radial acc^{ln} m/sec^3; R = radius of circular arc m. (Note mixed units.)
  In our example, therefore

$$L = \frac{100^3}{46.7 \times 0.3 \times 710}$$

$$= 100.532 \text{ m}$$

Stage 7
Calculate the shift.

Note
The shift of the curve is the distance between the line of the curve and the line of a similar curve having the same radius and joining the same straights without transitions. It is given by the formula

$$S = \frac{L^2}{24R}$$

∴ in our example

$$S = \frac{(100.532)^2}{24 \times 720}$$

$$= 0.585 \text{ m}.$$

Stage 8
Calculate the length along the tangent from the I.P. to the start of the transition.
  This is another formula

$$IT = \frac{L}{2} + (R + S) \tan \frac{\theta}{2}$$

$$= \frac{100.532}{2} + 220.585 \tan \frac{41°34'48''}{2}$$

$$= 323.847 \text{ m}$$

This enables us to identify the points at which the straights join the transitions by measuring back along the straight from the I.P.

Stage 9
We can now determine any point on the transition by using the equation of the curve.

The most useful form of the equation for design purposes is the cubic parabola

$$x = \frac{y^3}{6RL}$$

This can be used to generate a series of offsets, x, at any distance y along the straight from the beginning of the transition.

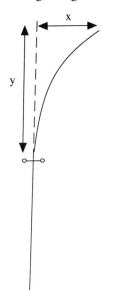

In order to plot the curve we might wish to generate a series of offsets at 10 m intervals:

y	y^3	$\frac{y^3}{6RL} = x$
10	1,000	0.002
20	8,000	0.018
30	27,000	0.062
		etc.

If we want to find the point at which the transition becomes the circular arc we can substitute the value of L for y.

In this case

$$x = \frac{L^3}{6RL} = \frac{L^2}{6R}$$

$$= \frac{(100.532)^2}{6 \times 720}$$

$$= 2.340$$

We have now fixed two points on the circular arc and we know its radius. Thus we have fixed the circle.

Stage 10 Setting Out

With the aid of a computer it is a fairly simple matter to rotate axes and move the origin, and hence calculate the co-ordinates of any point relative to a setting out grid. Working manually it may be desirable to set out using a theodolite set up at the start of the transition. The equation

$$\phi^2 = \frac{l^2}{2RL}$$

allows us to calculate a series of angles ϕ set off the line of the straight to find points on the transition at distances l from its start

l	$\dfrac{l^2}{2RL} = \phi$ rads	ϕ
10	0.0007	0°0′02″
20	0.0027	0°0′09″
30	0.0062	0°0′22″
		etc.

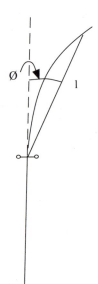

Appendix C

Example of vertical alignment and application of superelevation

Note

The fixing of levels for a highway will be undertaken in two stages: the design of the vertical alignment of a master reference line, such as the centre line and then the fixing of channel levels by addition or subtraction of the crossfall or superelevation required. Since the vertical alignment will have to take account of existing ground levels it is necessary to have a provisional horizontal alignment before commencing design of the vertical alignment.

The vertical alignment is made up of a series of straight gradients joined by vertical curves, which are normally parabolic. They can be 'hump' or 'sag' curves, depending on the gradients they join. The length of the curve, and hence its geometry, depends upon:

(a) The change in gradient to be accommodated by the curve.
(b) Whether the curve is 'hump' or 'sag'.
(c) The design speed.
(d) A 'K' value, specified by DTp in TD9/81 and determined from the required sight distance.

Problem

Design the vertical alignment for a section of road joining a gradient of +3% and one of −2% and having a design speed of 100 km/h.

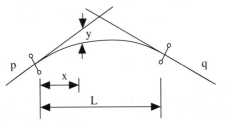

+3%
(1 in 33.3)

−2%
(1 in 50)

Note sign convention. Opposite slopes have opposite signs.

Stage 1
Determine the length of the vertical curve.
Required length is given by

$$L = K (p - q)$$

where K is 'K' value given in Table 3 TD9/81; p, q are gradients, expressed as percentages with appropriate signs.

Table 3, TD9/81 gives three alternative K values of 400, 105 or 59 according to whether we want Full Overtaking Sight Distance (FOSD), desirable minimum crest value or absolute minimum crest value. The choice will depend on feasibility and cost.

For this example we will adopt desirable minimum standards i.e. K = 105.

∴ length of vertical curve

$$L = K (p - q)$$
$$= 105 (3 - [- 2])$$
$$= 525 \text{ m}$$

Stage 2
Compute levels along the vertical curve. A formula to determine the offset from the gradient to the curve is

$$y = \left(\frac{q - p}{2L} \right) x^2$$

where y is the vertical offset from the continuation of the gradient to the curve and x is the distance along the curve.

In our example

$$\text{for } x = 100$$

$$y = \frac{-0.05}{1050} \times 100^2$$

$$= -0.476$$

If we say the level at the start of the curve is 100.000.

∴ level on gradient 100 m along curve

$$= 100.000 + 100 \times 0.03$$

$$= 103.000$$

∴ level on curve

$$= 103.000 - 0.476$$

$$= 102.524$$

Stage 3

To check the arithmetic we can calculate the level at the far end of the curve in two ways.

Using our formula for x = 525

$$y = \frac{-0.05 \times 525^2}{1050}$$

$$= 13.125$$

Level on gradient continued is

$$100 + 0.03 \times 525 = 115.750$$

∴ level on curve

$$= 115.750 - 13.125$$

$$= 102.625$$

Level of point where gradients meet

$$= 100 + 0.03 \times 262.5$$

$$= 107.875$$

∴ level on second gradient at far end of curve

$$= 107.875 - 0.02 \times 262.5$$

$$= 102.625$$

which is the same as before.

Crossfall/superelevation

On horizontal curves superelevation must be applied in accordance with TD9/81. This requires that superelevation should be in accordance with the formula

$$e = \frac{0.353V^2}{R}$$

where e = superelevation %; V = design speed in km/h; R = radius in m.

Thus, if we wish to calculate superelevation to be applied on a curve of radius 1000 m with a design speed of 100 km/h

$$e = \frac{0.353 \times 100^2}{1000}$$

$$= 3.53\%$$

If we have a 7.3 m wide carriageway with 1.0 m edging strips.

Total width of pavement = 9.3 m

∴ fall across pavement

$$= 9.3 \times 0.0353$$

$$= 0.328 \text{ m}.$$

Index